SYMMETRIES IN SCIENCE IV

Biological and Biophysical Systems

SYMMETRIES IN SCIENCE IV

Biological and Biophysical Systems

Edited by

Bruno Gruber and **John H. Yopp**

Southern Illinois University
Carbondale, Illinois

PLENUM PRESS · NEW YORK AND LONDON

Library of Congress Cataloging in Publication Data

Symmetries in science IV: biological and biophysical systems / edited by Bruno Gruber and John H. Yopp.

p. cm.

"Proceedings of a symposium entitled Symmetries in science IV: biological and biophysical systems, held July 24-27, 1989, at Landesbildungszentrum Schloss Hofen, Lochau, Vorarlberg, Austria"—T.p. verso.

Includes bibliographical references.

ISBN-13:978-1-4612-7884-9 e-ISBN-13:978-1-4613-0597-2

DOI: 10.1007/978-1-4613-0597-2

1. Symmetry (Biology)—Congresses. 2. Molecular biology—Congresses. I. Gruber, Bruno, 1936- . II. Yopp, John H.

QH506.S94 1990 90-6766

574—dc20 CIP

The articles "Symmetry in Synthetic and Natural Peptides," by E. Benedetti *et al.*, "Biomolecular Handedness and the Weak Interaction," by A. J. MacDermott and G. E. Tranter, and "Some Ideas on Evolution and Symmetry Breaking of Biosphere on Earth," by W. Thiemann relate to Department of the Navy Grant N00014-89-J-9043 issued by the Office of Naval Research. The United States government has a royalty-free license throughout the world in all copyrightable material contained therein.

Proceedings of a symposium entitled Symmetries in Science IV: Biological and Biophysical Systems, held July 24-27, 1989, at Landesbildungszentrum Schloss Hofen, Lochau, Vorarlberg, Austria

Softcover reprint of the hardcover 1st edition 1990

A Division of Plenum Publishing Corporation
233 Spring Street, New York, N.Y. 10013

FOREWORD

The symposium, "Symmetries in Science IV: Biological and Biophysical Systems," was held at Schloss Hofen, Vorarlberg, during the period July 24-27, 1989. Its purpose was to promote interaction between scientists working in the area of biological-biophysical research with scientists working in physics, mathematical physics, and mathematics.

Reviews in the field of biological-biophysical research were presented by the participants, and subsequently these presentations were analyzed by the interdisciplinary group in a workshop and by means of round table discussions. This volume contains the review presentations as a means of making this subject available to the general scientific community.

The symposium was co-sponsored by:

Southern Illinois University at Carbondale
Land Vorarlberg
U.S. Office of Naval Research, London

We wish to thank these institutions for their support. Moreover, we wish to thank the following individuals:

Dr. John C. Guyon, President
Southern Illinois University at Carbondale

Dr. Benjamin A. Shepherd
Vice President for Academic Affairs and Research
Southern Illinois University at Carbondale

Dr. Martin Purtscher
Landeshauptmann des Landes Vorarlberg

Dr. Guntram Lins
Landesrat des Landes Vorarlberg

Dr. Hubert Regner
Vorstand der Abt. Volksausbildung und Wissenschaft

Their generous support and steady encouragement has made Symposium IV possible.

Bruno Gruber
John H. Yopp

CONTENTS

SYMMETRY IN SYNTHETIC AND NATURAL PEPTIDES

E. BENEDETTI, B. DI BLASIO, A. LOMBARDI, V. PAVONE
and C. PEDONE

Department of Chemistry University of Napoli
Via Mezzocannone, 4 80134 Napoli - Italy

Peptides of synthetic or natural origin are compounds able to exert a variety of biological function: they are hormones, protein substrates and inhibitors, sweeteners, opioides, antibiotics, releasing factors, regulators of biological functions, citoprotectors and so on.

From the chemical point of view peptides are formed when two or more amino acid residues of the type shown above are condensed together, leading to a peptide unit (a secondary amide bond) and the formation of a dipeptide:

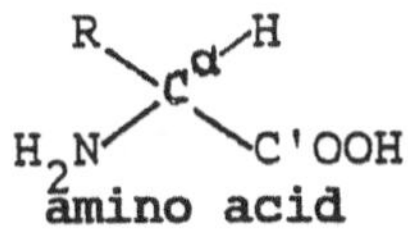

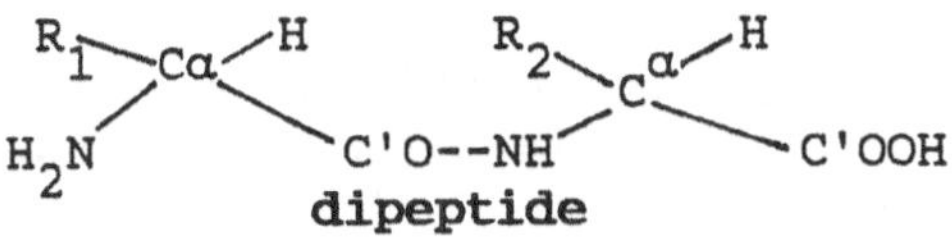

Peptides are, then, chains of a certain number of covalently linked amino acid residues each of which is intrinsically asymmetric, because of the optically active α-carbon atoms. The amino acid sequence along the chain, the spatial configuration of the asymmetric C^{α} atoms of each residue, the local conformation of part of the molecule or the overall conformation of the entire peptide, together with the intramolecular and intermolecular interactions of various types, are all important factors in determining the biological activity and the mechanism of its action.

The structural characteristic of the peptide unit formed through the linkage of residues i and i+1 are fully described in terms of geometry of bonds, conformation and non-planarity of chemical bonds. According to the IUPAC-IUB Commission on Biochemical Nomenclature (1), the complete description of the conformations of the backbone and of the side chains in peptide and proteins is that given in Figure 1.

Symmetries in Science IV
Edited by B. Gruber and J. H. Yopp
Plenum Press, New York, 1990

Accurate values for the geometry of the peptide unit as well as for that of most side chains of the twenty naturally occurring amino acid residues have been obtained by the analysis of literature data, using accurately determined crystal structures of small and medium size peptides (2). The magnitudes of bond lengths and valence angles are generally very constant and impervious to intermolecular interactions.

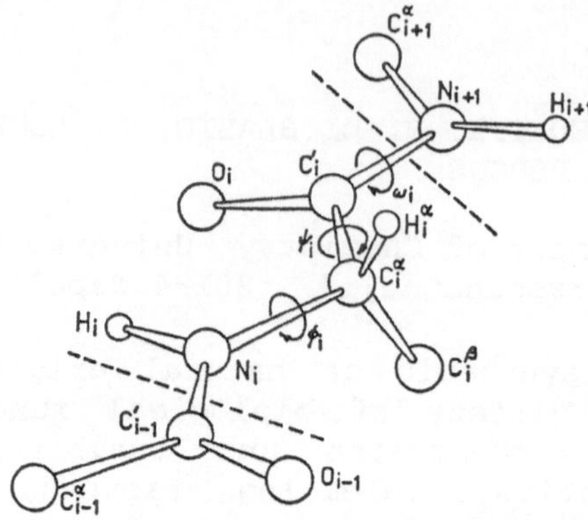

FIG. 1. Two peptide units in a polypeptide chain in a fully extended conformation ($\varphi_i = \psi_i = \omega_i = 180°$) for L-residues. The limits of a residue are indicated as dashed lines and the reccomended notation for atoms and conformational angles are shown.

Molecular conformations is most conveniently and precisely characterized by torsion angles. The shape and consequently the symmetry or the asymmetry of a particular peptide molecule is the consequence of a certain succession of torsion angles $\varphi_i, \psi_i, \omega_i$, for each amino acid residue, while the torsion angles χ_i fully describe the side chain conformation. The ω_i angles in linear peptides usually present values close to 180° corresponding to a trans arrangement of the type:

which is energetically more stable than the cis arrangement ($\omega_i \approx 0°$) by about 2 Kcal/mol (3). Of course a cis conformation is a necessity in small cyclic peptides (di- and tripeptides) but in higher cyclic peptides this arrangement may or may not occur; in any case the occurrence of a cis peptide unit in a linear chain is to be considered a very rare event. Consequently the molecular conformation of a peptide can be visualized in the (φ, ψ) space by the energy of the molecule as function of these torsion angles. The maps obtained show that all best known secondary structures assumed by a peptide chain, such as the β-structure, the α-helix, the 3_{10}-helix, occur within energy minima. Conformational energy contour maps have been accurately calculated by Zimmerman et al. (4) and Figure 2 shows the maps obtained for glycine and alanine.

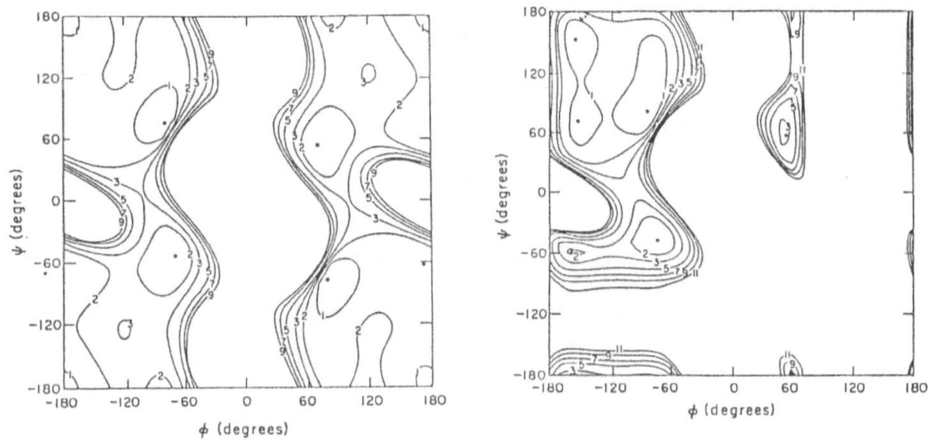

FIG. 2. Conformational energy contour maps of N-acetyl-N'-methyl-glycineamide (on the left) and of N-acetyl-N'-methyl-alanineamide (on the right).

The symmetry shown in the Gly map disappears for the optically active Ala residue and the conformational space available to a Gly residue (54 % of total area) is greatly reduced by the substitution of the hydrogen atom by a methyl group in the Ala residue (18 % of total area).

Thus, very often peptides do not show any symmetry and the only cases for which certain elements of symmetry are maintained are those concerning cyclic peptides or linear peptides in helical arrangement. In the following some examples of symmetry in cyclic and linear peptides will be presented, keeping in mind that any element of symmetry, such as mirror planes or center of inversion, which invert the configuration of the optically active α-carbon atom, should be banned, since it would change the chemical nature of the peptide (unless such inversion of configuration is constitutionally present in the peptide as in regularly alternating L,D peptides). Furthermore, rather often, in the literature the symmetry reported for a peptide refers only to the backbone atoms of the molecule, while side chain atoms are not considered.

CYCLIC PEPTIDES

The structure of cyclic peptides is simplified because the allowed conformations are reduced by the cyclic character of the molecules, which are constrained to contain bends in the backbone.

The simplest cyclic system is represented by cyclic dipeptides. Many natural products contain elements of this structural system, in which the backbone atoms forms a six membered ring. Because of the presence in the backbone of different atomic species having different hybridization, only the center of inversion and symmetry axis perpedicular to the dominant plane of the ring need to be considered in order to define ring conformation in cyclic dipeptides. The conformation experimentally observed for cyclic dipeptides can be grouped in quasi-planar conformation, boat conformation with C^{β} atoms in axial or in equatorial position and chair conformation, as shown in Figure 3.

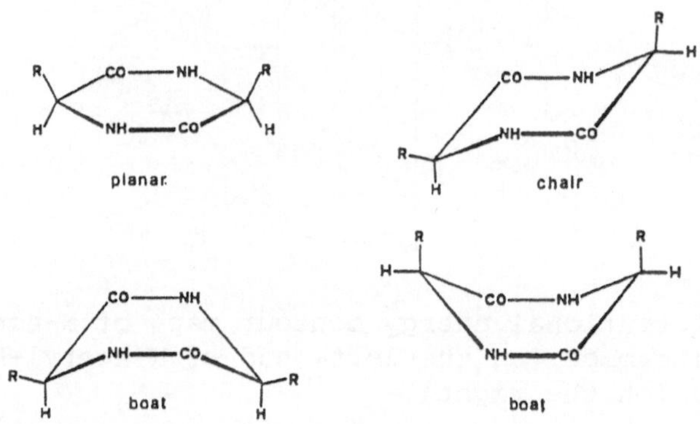

FIG. 3. The four possible conformations for cyclic dipeptides

Quasi-planar and boat conformation have a two-fold symmetry element perpendicular to the average plane of the ring, while chair conformations have no symmetry element. The center of inversion is shown in the cyclic dipeptide of glycine and possibly in cyclic dipeptides of residues with opposite configurations.

The four conformations are characterized by a different sequence of the conformational angles $\varphi_1, \psi_1, \omega_1, \varphi_2, \psi_2, \omega_2$. Because of the steric requirements, the two dipeptide units are forced to be cis ($\omega \approx 0°$) with only small deviations from planarity. Quasi-planar conformations present values for φ_1, ψ_1, φ_2, ψ_2 close to zero; boat conformations with axial substituents on the C^{α} atom are characterized by positive φ values and negative ψ values, while boat conformations with equatorial substituents on the C^{α} atom have negative φ values and positive ψ values. The chair conformation, finally, is characterized by an alternating sequence of positive and negative values of the six conformational angles.

For cyclic tripeptides, if one does not take into account side-chain atoms, the solid state conformation (5-7), confirmed by solution results also, shows an approximate three-fold symmetry axis perpendicular to the mean plane of the backbone ring atoms. The peptide units are all cis ($\omega \approx 0°$) and the φ and ψ conformational angles present values ranging from -90° to -110° for φ and 80° to 105° for ψ, for peptides constituted only by L residues.

Several evidences have been gathered on the remarkable similarity shown by cyclic tetrapeptides (2,8-10): their conformational features are strikingly similar in spite of the difference in the chemical structure and confirm solution NMR spectroscopy data on the existence of a predominant conformation. The similarity of the ring symmetry should lie in the intrinsic conformation of the peptide chain itself, which shows for the atoms of the ring an approximate center of inversion (i symmetry).

The flexibility of the ring and consequently the possibility of observing multiple conformations increases with increasing number of residues in the ring system. Then, intraring or transannular interactions, such as hydrogen bonds between N-H and C=O groups, play an important role in stabilizing preferentially one conformation.

Among cyclic peptides, hexapeptides with their 18-membered rings are of special interest, since they are found quite frequently in many biologically important molecules. Literature data show that several significantly different conformations have been experimentally observed; accordingly hexapeptides can be broadly divided in two groups: 1) structures in which the backbone of the peptide chain nearly retains a symmetry element, like a center of symmetry or a two-fold axis (resulting in opposite or equal values, within about 25° to 30°, for the φ and ψ angles of residues i and i+3, respectively). Special cases are represented by cyclic hexapeptides with L,D residues, for which an higher symmetry has been experimentally seen in the solid state as well as in solution.

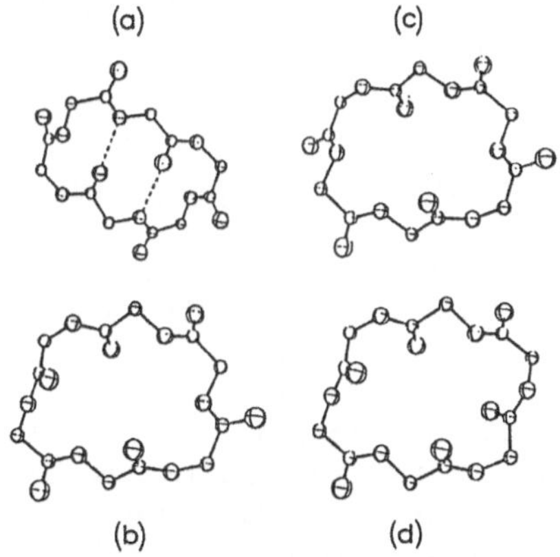

FIG. 4. Four different conformers of <u>cyclo</u>-$(Gly)_6$.

2) asymmetric backbone structure.

The best example which illustrate the above classification is represented by the solid state structure of cyclo-$(Gly)_6$ (11). For this peptide four different conformers, crystallized side by side in the same cell (Figure 4) are seen, demonstrating in general the flexibility of the ring system for hexapeptides, and furthermore, the existence of multiple conformers is indicative of a very little difference in the energy of such structures.

The most and the least populated conformers of cyclo-$(Gly)_6$ retain in the solid state a center of symmetry as a crystallographic element of symmetry.

The structures of cyclo-$(L\text{-Ala-L-Pro-D-Phe})_2$ (12), cyclo-$(Gly\text{-L-Pro-D-Phe})_2$ (13), cyclo-$(Gly\text{-L-Tyr-Gly})_2$ (14) and cyclo-$(L\text{-Pro-L-Val-D-Phe})_2$ (15) exhibit in the solid state either a crystallographic or an approximate C_2 symmetry, as shown in Figure 5).

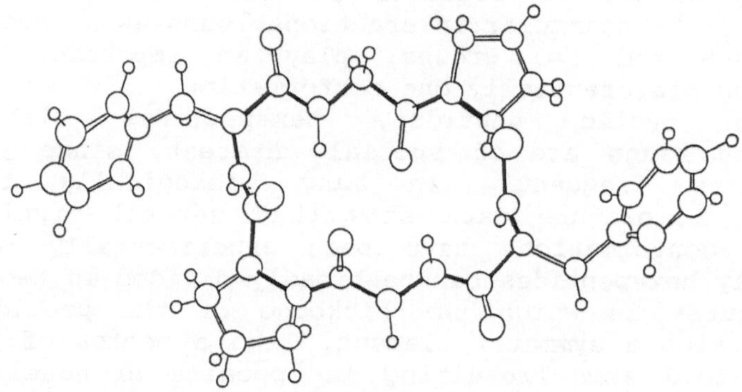

FIG. 5. Molecular structure of <u>cyclo</u>-$(Gly\text{-Pro-Phe})_2$.

Finally the structures (16) of cyclo-$(L\text{-Phe-D-Phe})_3$ and cyclo-$(L\text{-Val-D-Val})_3$, are examples of cyclic hexapeptides retaining a center of symmetry or presenting an higher symmetry (S_6 symmetry), respectively.

LINEAR PEPTIDES

A. Relevant Determinants of Secondary Structure

The secondary structure of linear peptides is commonly stabilized by short range atomic interactions in the molecule.

The most important stabilizing factors are the intramolecular hydrogen bonds of the N-H···O=C type. An H-bond between the donor group N-H of residue m and the acceptor group of the residue n is designated as an m--n H-bond. Thus in a system of four linked peptide units, shown in Figure 6, the possibilities of intramolecular H-bond are the 2→2 (or 3→3 or 4→4), the 2→3 (or 3→4), the 2→4, the 3→1 (or 4→2 or 5→3), the 4→1 (or 5→2) and the 5→1. The resulting intramolecular hydrogen-bonded conformations can be also characterized by the number of atoms in the ring obtained by the H-bond formation. Then the above mentioned conformations are called C_5, C_8, C_{11}, C_7, C_{10} and C_{13} conformations, respectively. Among these the more commonly found

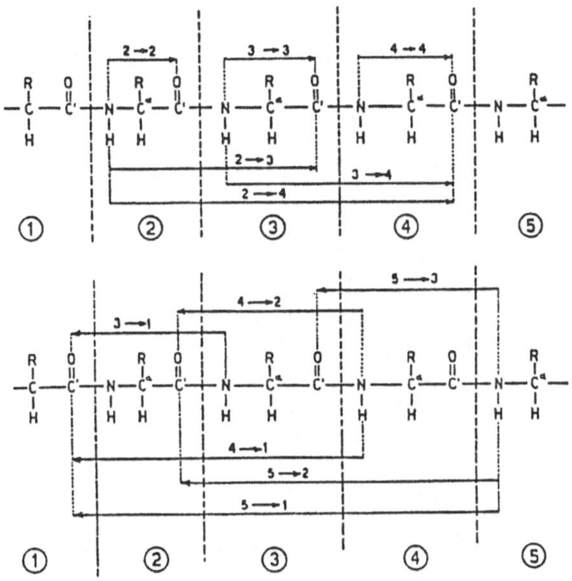

FIG. 6. Possible intramolecular hydrogen bonds occurring in a sistem of four linked peptide residues.

types are the C_{10} and C_{13} conformations. Fewer examples of C_5 and C_7 conformations have also been reported in the literature. With the exception of the 2→2, or C_5 conformation which consists of an intraresidue H-bond, giving rise to an extended conformation, the other types of bonded conformations imply a reversal in the polypeptide chain direction. Consequently an isolated C_7, C_{10} or C_{13} conformation along the peptide chain is also called chain reversal or γ-, β- or α-turn (17) respectively.

The C_5 conformation has been proposed and experimentally seen in a number of peptides (18,19). Most of them are, however, homopeptides derived from α,α-dialkylated glycine residues such as shown in Figure 7.

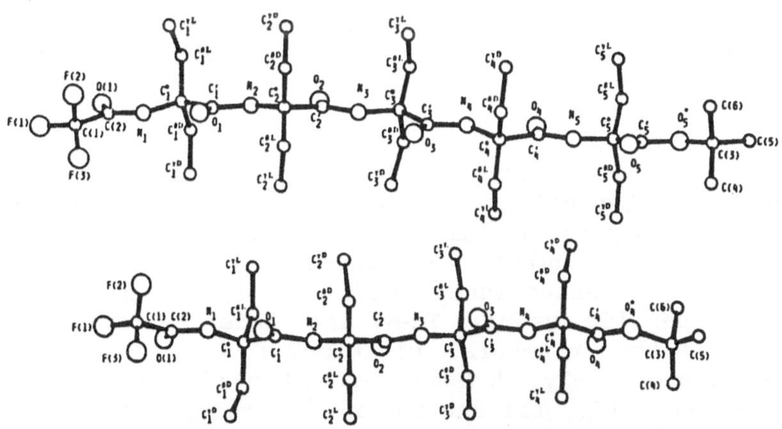

FIG. 7. Structure of TFA-(Deg)$_n$-O$\underline{t}$Bu with n = 4, 5.

Glycine or dialkylated glycine residues show the C_5 conformation more than other residues, while the dissymmetry introduced by the side chain consistently increases the warping of these residue.

For a 3→1 C_7 conformation or γ-turn, the planes of the peptide units, preceding and following the central residue, make an angle of about 125°. If the central residue has a chiral C^α atom then two different structures with equatorial and axial side-chains exist depending on the position of the C^α-C^β bond with respect to the intersection line of the two amide planes. For an L-residue at position 2 in an equatorial 3→1 intramolecularly hydrogen-bonded conformation, the torsion angles assume values φ_2=-75, ψ_2=+50 , while they assume opposite values for an axial position of the side chain. Both conformation have been experimentally observed in peptides (20-22).

Among the various intramolecularly hydrogen bonded conformations, the C_{10} forms are the most frequently found in either the solid state or in solution (2,17,23). This bond is characterized by the values of the φ and ψ conformational angles of the central residues at position i+1 and i+2. The formation of the H-bond between the N-H of residue 4 and C=O of residue 1 depends on the relative orientation of three peptides units. Consequently various types of orientations produce different β-bends or C_{10} conformations. Four "standard" β-bends, called type I, II_1, II_3 or III, have been characterized and their φ_{i+1}, ψ_{i+1}, φ_{i+2}, ψ_{i+2} are given in Table I.

Table I. **Dihedral Angles for the "Standard" β-bend of various type.**

β-bend type	φ_{i+1}	ψ_{i+1}	φ_{i+2}	ψ_{i+2}
I	-70	- 35	-70	+35
II_1	-60	+130	+85	+35
II_3	-60	+130	+85	-35
III	-70	- 35	-70	-35

Types I and III have similar φ and ψ values for the i+1 residue, while the values for the i+2 residue are more broadly distributed (they differ mainly by the changed orientation of the third peptide unit). Analogously two forms of type II β-bends can be distinguished, differing by the orientation of the third peptide group. They are labelled II_1 and II_3 according to the structural relations to type I and type III β-bends, respectively. β-Bends are then classified in eight groups (I, II_1, II_3, III and their mirror images).

The intramolecular hydrogen-bonded C_{13} conformation (5→1) occurring between the N-H of the 1+4 residue and the

C=O of the i residue is similar to that occurring in α-helices. In helical structures, however, there is a regular succession of the φ and ψ conformational angles, while in an isolated C_{13} ring structure (or α-bend) a larger flexibility can be observed with a larger variability for the three pairs of the φ, ψ conformational angles of the central residues (i+1, i+2, i+3). Most of the time the trans conformation of the two internal peptide units has been observed. Examples are found in the structures of biologically active peptides and depsipeptides such as β-amanitin (24), valinomycin (25-27), isoleucinomycin (28).

B Helical Structures

If the conformational angles φ, ψ assume the same values for all residues along a peptide chain, then an helical conformation is generated. In the following we will consider only helical structures stabilized by intramolecular H-bonds of the type discussed above.

Thus, we will consider helices in which 3→1, 4→1 and 5→1 H-bonds are the factor stabilizing the conformation, so that they can be visualized as an infinite succession of ring structures (C_7, C_{10}, C_{13}) of the type already discussed. A list of properties and parameters for these helices is given in Table II.

Table II. **Characteristic Parameters of Polypeptide Helical Structures.**

Parameters		2_1-helix	3_{10}-helix	α-helix
Symmetry		2_1	3_1	18_5
Residue Repeat (length per residue in A)		2.75	2.0	1.50
Type of Intramolecular H-bond		3→1	4→1	5→1
Number of Atom in Ring		7	10	13
Conformational Angles	φ	-80	-60	-55
for Right-Handed Helices	ψ	70	-30	-45
of L-Residues (in degrees)	ω	160	180	180
Designation according to Bragg		2_7	3_{10}	3.16_{13}

The 2_1-helices, obtained by a consecutive succession of seven-membered hydrogen-bonded ring structures of the 3--1 type can also be described in terms of symmetry through the operation of a two-fold axis followed by a translation along this axis. This helix, in the Bragg et al. notation (29) is designated as a 2_7 helix, where 2 indicates the symmetry of the helix and 7 is the number of atoms in the H-bonded ring. In the current nomenclature for helices (2_1-helix) 2 stands for the number of units to form a perfect repeat after 1 turn. In principle two helices of this type (left and right-handed) are possible for a polypeptide chain consisting of all L-residues. For them the position of the C^{β} side chain atom (at position 2 in each successive C_7-ring structure) is equatorial or axial for right- and left-handed helices, respectively. This structure, proposed long time ago, has never been observed in peptides or polypeptides and one of the reason could be the necessity for this helix of substantial systematic deviation from planarity of peptide units.

The 3_{10}-helix can be visualized as a structure consisting of a succession of ten-membered intramolecularly hydrogen-bonded type III β-bend for left-handed and III' β-bend for right-handed helices. Each residue in the peptide chain present values for φ and ψ conformational angles of (-60°, -30°) or (60°, 30°) for right- and left-handed helices, respectively. No substantial distortion from planarity for peptide units seems to be required. In the 3_{10}-helix, 3 residues per turn give rise to an exact three fold symmetry with hydrogen bonds oriented nearly parallel to the triad axis. Peptides containing high percentage of the α,α-dialkylated α-amino <u>iso</u>butyric acid (Aib) (or dimethylglycine), a residue occurring in many microbial peptides, show a structure consisting of successive multiple type III β-bends resulting in incipient 3_{10}-helix (30). Peptides, containing Aib residues, show a very high propensity to fold into 3_{10}- or α-helices with both right- and left-handed twists of the chain.

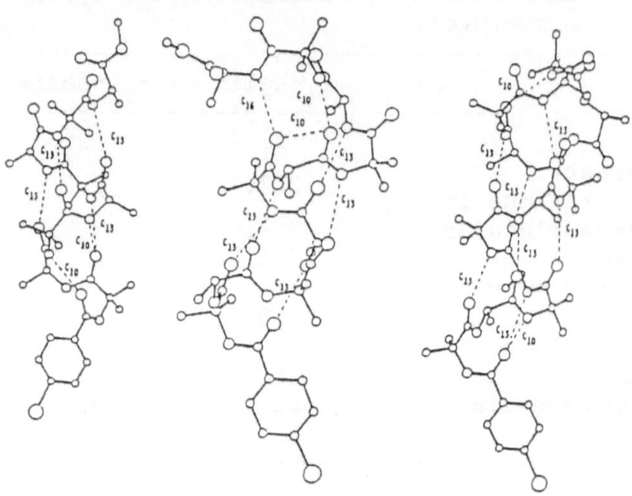

FIG. 8. Structures of Z-(L-Ala-Aib)$_n$-OMe with n = 4, 5, 6.

Among the possible helical structures that peptides and polypeptides can assume, the α-helix is by far the best know. The structure consists of eighteen peptide residues in five turns with all donors (N-H) and acceptors (C'=O) of hydrogen bonds nearly parallel to the helix axis. Successive 5→1 (C_{13}) intramolecularly H-bonds stabilize the structure. According to Bragg et al. (29) the correct nomenclature of the α-helix should be 3.6_{13} (3.60 residues per turn in an H-bond 13-membered ring structure). The helix is, thus, generated with a screw operation by a rotation of 100° and an axial translation of 1.50 A for successive peptide units. The α-helix can be either right- or left-handed. It is the most stable conformation for polyglycine, where it exist in two isoenergetic enantiomorphs, while for other homopolypeptides, having substitution on the C^{α} atom, the right-handed form is slightly (≈ 0.2 Kcal/mol) more stable that the left-handed one. Long segments in protein crystal structures are often found in α-helical conformation; few peptides have also be reported to exist in the solid state, as well as in solution, as α-helices. Among these, peptides of the series Boc-(L-Ala-Aib)$_n$-OMe (30) with n = 5 or 6 and Boc-Ala-(Aib--Ala)$_2$-Glu-(OBzl)-Ala-(Aib-Ala)$_2$-OMe (31) show most of the residues in the α-helical conformation, as seen in Figure 8.

In the last decade peptides and polypeptides with a regular sequence of enantiomeric residues (L and D) along the chain have received considerable attention (32-35), because their conformational behavior somehow differs from that of homoconfigurational peptides or polypeptides. In fact, these molecules could assume not only know structures, such as the α-helix but also more specific structures different from those characterizing poly-(α-amino acid) chains. Most of the interest in the conformation of polypeptides with regularly alternating L- and D- amino acid residues derives from the search of the possible structures for the natural transmembrane ion carrier antibiotic, gramicidin A, a linear pentadecapeptide with regular alternation of L- and D- residues. On the basis of NMR and spectroscopic results, a completely new class of helical structures has been proposed (32) in which enantiomorphous residues in β conformation alternate along the chain. Several type of helices, called $\beta_{L,D}$ where n is the number residues per turn, may be possible. Double helical structures (33) with both parallel or antiparallel strands denoted as

$$\uparrow\uparrow \beta\beta^{n}_{L,D} \quad \text{and} \quad \uparrow\downarrow \beta\beta^{n}_{L,D}$$

respectively have, also, been proposed (Figure 9).

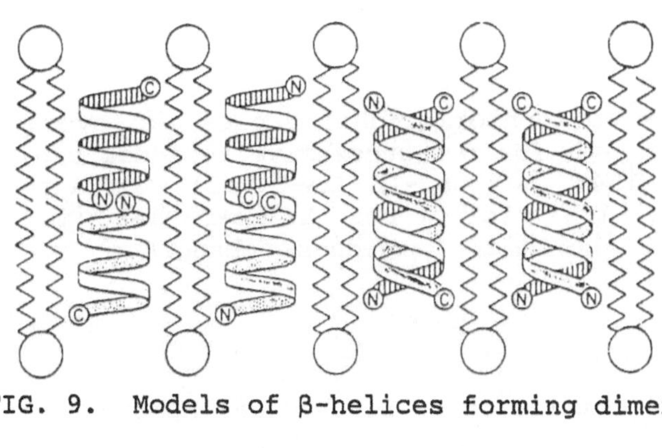

FIG. 9. Models of β-helices forming dimers.

Both families of single stranded $\beta^{n}_{L,D}$ and double-stranded $\beta\beta^{n}_{L,D}$ helices are stabilized by intrachain or interchain hydrogen bond, respectively.

Recently, detailed structural and conformational parameters for two double stranded ββ helices, as found in the crystals of the two octapeptides t-Boc-(L-Val-D-Val)$_4$-OMe (36,37) and t-Boc-(L-Phe-D-Phe)$_4$-OMe (38), have been reported, Figure 10. The double stranded antiparallel left-handed helix of t-Boc-(L-Val-D-Val)$_4$-OMe is build up through the operation of a crystallographic binary axis running perpendicular to the chain axis of one octapeptide molecule. The two octapeptide chains wind up around each other giving rise to a ↑↓ $\beta\beta^{5,6}_{L,D}$ double helix. The two independent octapeptide chains of t-Boc-(L-Phe-D-Phe)$_4$-OMe are analogously associated by intrachain hydrogen bonds with the formation of a double stranded antiparallel right-handed ↑↓ $\beta\beta^{5,6}_{L,D}$ helix. In this case no symmetry relate the two chains and the handedness of the dimeric helix is opposite to that found in t-Boc-(L-Val-D-Val)$_4$-OMe.

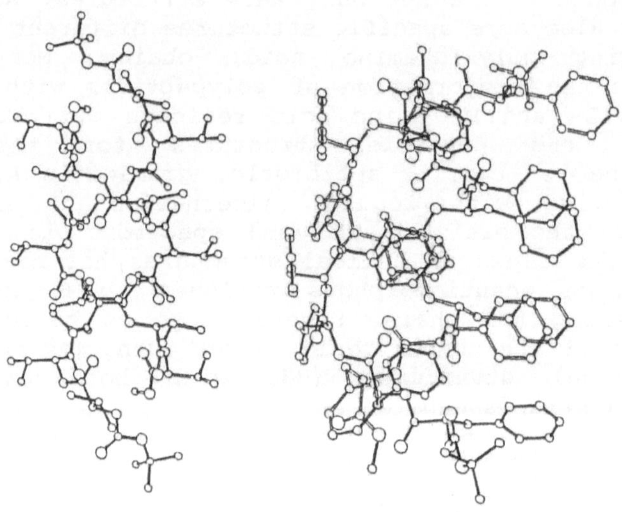

FIG. 10. Structures of t-Boc-(L-Val-D-Val)$_4$-OMe (left) and of t-Boc-(L-Phe-D-Phe)$_4$-OMe (right).

In terms of conformational angles the average values of φ_L, ψ_L and φ_D, ψ_D for two successive residues are -138°, +107° and +109°, -154° in t-Boc-(L-Val-D-Val)$_4$-OMe and -109°, +154° and +152°, -153° for t-Boc-(L-Phe-D-Phe)$_4$-OMe, since the inversion of the helix handedness changes $\varphi_L = -\varphi_D$ and $\psi_L = -\psi_D$.

REFERENCES

1. IUPAC-IUB Commission on Biochemical Nomenclature,

Biochemistry, 9, 3421 (1970) and **Biochemistry**, 11, 1726 (1972).
2. Benedetti, E., in "**Chemistry and Biochemistry of Amino Acids, Peptides and Protein**", Ed. B. Weinstein, Ed., Vol. 6, p. 105.
3. Ramachandran, G. N. and Sasisekharan, V., **Advan. Protein Chem.**, 23, 287 (1968).
4. Zimmerman, S. S., Pottle, M. S., Nemethy, G. and Scheraga, H. A., **Macromolecules**, 10, 1 (1977).
5. Kartha, G., Ambady, G. and Shankar, P. V., **Nature**, 247, 204 (1974).
6. Kartha, G. and Ambady, G., **Acta Crystallogr.**, B31, 2035 (1975).
7. Groth, P., **Acta Chem. Scand.**, A30, 838 (1976).
8. Decleracq, J. P., Germain, G., van Meerssche, M, Debaerdemackar, T., Dale, J. and Titlestad, K., **Bull. Soc. Chim. Belg.**, 84, 275 (1975).
9. Groth, P., **Acta Chem. Scand.**, 24, 780 (1970).
10. Flippen, J. L. and Karle, I. L., **Biopolymers**, 15, 1081 (1976).
11. Karle, I. L. and Karle, J., **Acta Crystallogr.**, 16, 969 (1963).
12. Brown, J. N. and Teller, R. G., **J. Am. Chem. Soc.**, 98, 7565 (1976).
13. Brown. J. N. and Yang, C. H., **J. Am. Chem. Soc.**, 101, 445 (1979).
14. Ramachandran, G. N. and Shamala, N., **Acta Crystallogr.** A32, 1008 (1976).
15. Flippen-Anderson, J. L., in "**Peptides: Structure and Biological Function**" E. Gross and J. Meinhofers, eds. Pierce Chem. Co., Rockford 1979 p.145
16. Pavone, V., Benedetti, E., Di Blasio, B., Lombardi, A., Pedone, C., Tomasic, L. and Lorenzi, G. P., **Biopolymers**, 28, 215 (1989).
17. Toniolo, C., **CRC Crit. Rev. Biochem.**, 9, 1 (1980)
18. Benedetti, E., Palumbo, M., Bonora, G. M. and Toniolo, C., **Macromolecules**, 9, 417 (1976).
19. Benedetti, E., Barone, V., Bavoso, A., Di Blasio, B., Lelj, F., Pavone, V., Pedone, C., Bonora, G. M., Toniolo, C., Leplawy, M. T., Kaczmarek, K. & Redlinski, A., **Biopolymers**, 27, 357, (1988).
20. Flippen, J. L. and Karle, I. L., **Biopolymers**, 15, 1081 (1976).
21. Karle, I. L., in "**Peptides**", M. Goodman and J. Meienhofers, eds., J. Wiley and Sons, New York, p. 274 (1977).
22. Karle, I. L., in "**Perspectives in Peptides Chemistry**", A. Earle, R. Geiger and T. Wieland, eds., S. Kargel, Basel, p. 261 (1981).
23. Nemethy, G. and Scheraga, H. A., **Q. Rev. Biophys.**, 10, 239 (1977).
24. KostanseK, E. C., Lipscomb, W. N., Yocum, R. R. and Thiessen, W. E., **Biochemistry**, 17, 3790 (1978).

25. Duax, W. L., Hauptman, H., Weeks, C. H. and Norton, D. A., **Science**, 176, 911 (1972).
26. Smith, G. D., Duax, W. L., Langs, D. A., De Titta, G. T., Edwards, J. W., Rohrer, D. C. and Wheeks, C. M., **J. Am. Chem. Soc.**, 97, 7242 (1975).
27. Karle, I. L., **J. Am. Chem. Soc.**, 97, 4379 (1975).
28. Pletnev, Y. Z., Galitskii, N. M., Smith, G. D., Weeks C. M. and Duax, W. L., **Biopolymers**, 19, 1517 (1980).
29. Bragg, W. L., Kendrew, J. C. and Perutz, M. F., **Proc. Roy. Soc. London**, A203, 321 (1950).
30. Benedetti, E., Di Blasio, B., Pavone, V. and Pedone, C., **Second Forum on Peptides**. Eds. A. Aubry, M. Marraud, B. Vitoux. Colloque INSERM/John Liddey Eurotext Ltd., vol. 174, p. 27 (1989).
31. Winter, W., Butters, J., Hutter, H., Pauls, N., Sheldrick, G. M., Schimdt, H.and Jung, G., **Proc. of the Seventh American Peptides Symp., Madison, Wisconsin**, (1981) Abstract Q8.
32. Urry, D. W., **Proc. Natl. Acad. Sci. U.S.A.**, 68, 672 (1971).
33. Veatch, W. R., Fossel, E. T. and Blout, E. R., **Biochemistry**, 13, 5249 (1974).
34. Lotz, B., Colonna-Cesari, F., Heitz, F. and Spach, G., **J. Mol. Biol.**, 106, 915 (1976).
35. Lorenzi, G. P. and Tomasic, L., **J. Am. Chem. Soc.**, 99, 8322 (1977).
36. Benedetti, E., Di Blasio, B., Pedone, C., Lorenzi, G. P., Tomasic, L. and Graulich, B., **Nature**, 282, 619 (1979).
37. Di Blasio, B., Benedetti, E., Pavone, V., Pedone, C., Spiniello, O. and Lorenzi, G. P., **Biopolymers**, 28, 193 (1989).
38. Di Blasio, B., Benedetti, E., Pavone, V., Pedone, C. and Lorenzi, G. P., **Biopolymers**, 28, 203 (1989).

LOCAL AND GLOBAL MODEL OF FIBONACCI PLANT SYMMETRIES

Szaniszló Bérczi

Eötvös Loránd University
H-1088 Budapest, Rákóxzi u.5. Hungary

Introduction

In this review paper the structural modelling of Fibonacci plant symmetries will be divided to three hierarchy levels: to unit level, to structural family level and to structural class level. On the other hand models arguing from local interactions will be distinguished and groupped from those models which attack the problem by the global characteristics of the Fibonacci plant structure.

Distinction between local and global comes from crystallography. There the classical concept was that local order is reflected by the global symmetric appearance of euhedral crystals. The hierarchical division comes from system theory. There levels of operations, rules, structure form a dimension - it may be called dimension of embedding - in the description and modelling of a problem.

According to the local and global, and the hierarchical classification the following 2x3 matrix is the map of this review:

	LOCAL MODELS	GLOBAL MODELS
Hierarchy	usually golden angle from neighbourhood relations /one parameter/	band structure and global lattice relations /two parameters/
UNIT i.e. a plant	golden angle in phyllotaxis D'Arcy Thompson, Coxeter	two harms of $r \sim e^{\varphi}$ curves in Fibonacci numbers
STRUCTURAL FAMILY i.e. Fibonacci plants	Adler, Gierer-Meinhardt, cellular automaton model /Bérczi,1985/	physico-topological model system /band sliding + initial mirror symmetry/Bérczi, 1976, 1978, 1987/
STRUCTURAL CLASS	phyllotaxis classes Richter-Schranner	irreducible representation of deformation equivalent lattices on cylinder /Bérczi,1988/

Fig.1

Symmetries in Science IV
Edited by B. Gruber and J. H. Yopp
Plenum Press, New York, 1990

Program

In this framework of Fig.1. our interest will be focused to a physical, topological, global model and a related cellular automaton model. They embrace the three hierarchy levels in the global region /Bérczi, 1976, 1978, 1985, 1989./.

The central model is physical in the sense of kynematical description: one /alternating/ operation connect family members which start at the simplest case, which serves as initial condition. The model system is topological in the sense, that topological transformations serve 1. preserving structure during normalization on unite level, and 2. changing Fibonacci structure between family members by a regular neighbourhood-changing operation. It is global in the sense that always a global subsystem /the band/ is the object of transformations.

In our program the following transformations and operations will be carried out in the three hierarchic levels:

level	transformation or operation	result
UNIT	equivalency-classification and normalization of Fibonacci plant structures	normalized representatives
FAMILY	ordering the square-lattice normalized representatives according to their growing Fibonacci-numbers structure	ordered normalized representatives
	Genetic deduction sequence of normalized representatives by the part-band-sliding operation	family-tree of normalized representatives
CLASS	reduction of any cylindrical lattice by part-band-sliding together operation	irreducible representatives
	classification of cylindrical lattice systems by their irreducible representatives	deformation-equivalent cylindrical class

Unit level transformations

Fibonacci plants are plants that have globally rotational symmetric subsystems /organs/, where the surface lattice system of repeating congruent /or similar/ elements /"cells"/ form bands, and the number of these bands winding to the left and to the right are two neighbouring Fibonacci numbers. The bands suggest a frame of reference for lattice elements which are numbered according to their heights on the stem or organ /which has rotational axis perpendicular to the water-level/.

At unit level a transformation is necessary to correspond each other the two different structural appearances of Fibonacci plants. One is phyllotaxis, the other is the dense mosaic-lattice on organs, most frequently on the tips and crops of

Fig.2. Global characteristics of Fibonacci plant structures on unit level. a. rotational symmetry, b. constant angle of meridian crossing of edges or lattice lines, c. both cylindrical and discoidal appearance, d. two opposite winding band-systems.

plants. From a structural point of view, phyllotaxis can be considered as a dot system variant of a Fibonacci structure stretched on to the stems of plants. A simple Dirichlet transformation can show that the grid points of phyllotaxis have essentially the same structure as a dense mosaic-lattice has.+

+According to Dirichlet's method, cells are formed around dots by drawing the lines which are perpendicular to the lines connecting two dots and are halfway between the two dots.

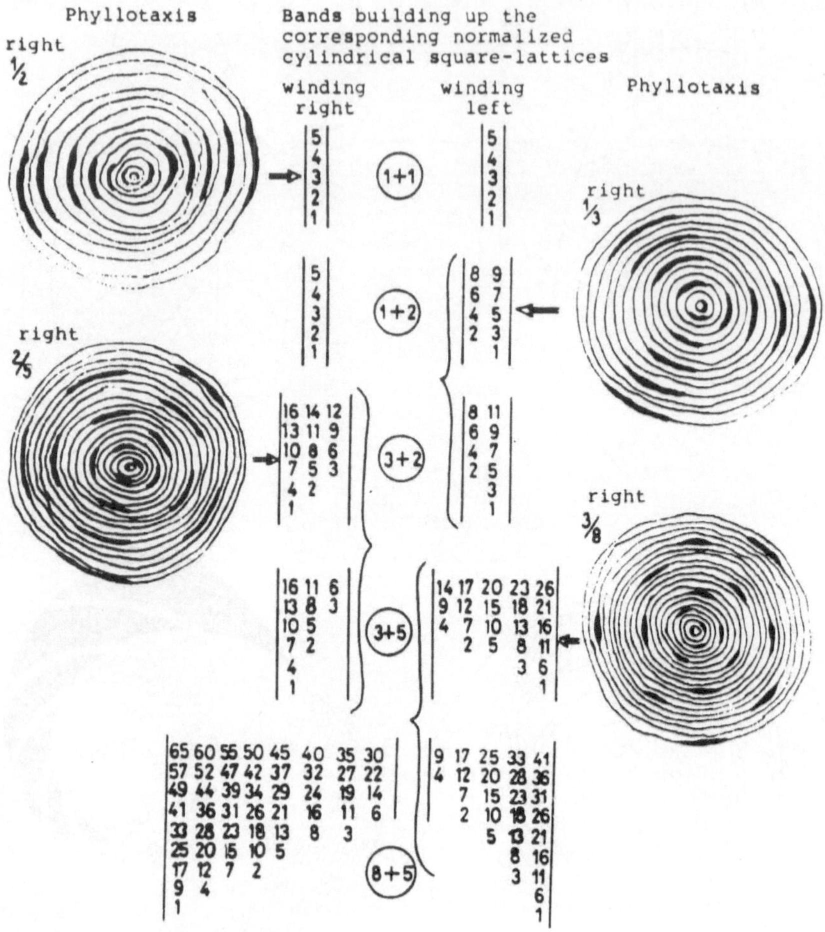

Fig. 3. Dirichlet and topological transformations of Fibonacci plant organs give into our hands the normalized representatives of different phyllotactic and dense-lattice-mosaic structures. Comparison of bands connected by brackets /one above the other/ show to recognize the part-band sliding operation, which builds Fibonacci plant structures into a family.

Another characteristic, a boundary condition, is also important to the normalization of different Fibonacci plant structures. This boundary condition is related to the rotational symmetry. Unordered directions around the developing organ result in this global ratational /cylindrical/ symmetry of its structure. On this rotational symmetrical surface the lattices of Fibonacci plants have a characteristic property. The curves of the lattice system cross the meridians of the surface at constant angles. This invariant property makes it possible that using a topological transformation any kind of Fibonacci-lattice can be mapped onto the square grid lattice /or equivalently, to a smooth cylindrical lattice/ without losing the main character of the Fibonacci structure./Fig.3./

The family level connecting operation and the physico-topological model system

On family level the common structural invariants of all Fibanacci plants are investigated. We can find these common invariants in two things belonging together. One is a structure rearranging lattice deformation operator, the other is the mirror symmetric initial condition. Mirror symmetry is so preserved that two enantiomorphic families of Fibonacci plants exist.

The normalized Fibonacci lattices were separated to their two band system components and these were arranged according to the increasing Fibonacci numbers, one above the other in Fig.3. Comparing bands with equal width, one above the other /connected by brackets/ we can see that they are different in relation to the position of the constituent part-bands. Part-bands in the lower Fibonacci lattice are displaced one lattice unit in one direction if we compare them to part-bands of the upper Fibonacci lattice. On the basis of this recognition we can construct an operation which connects and deduces all the family of Fibonacci plant structures. The operation is the part-band sliding /displacement/ which should be used alternately in the two lattice directions. The operation must start from the simplest case, the one+one band arrangement, which is well known as the structure of the wheatear. Continuing the sliding movement according to the alternating rule we can build up the structure of any higher Fibonacci plants. /Fig.4./

We could see that together with the part-band sliding operation the mirror symmetric outset also belongs to the common properties of Fibonacci plant structures. Because Fibonacci plant structures may represent a macroscopic final stage of a deformational development of Fibonacci lattices, they can be considered as preservers of initial mirror symmetry /although in a hidden form/.

But the Fibonacci structure is in direct logical relation to the mirror symmetric initial condition in an other, visible sense, too. At the first step of sliding operation the initial mirror symmetry is violated. This way two enantiomorphic families of Fibonacci plants may begin to develop. Develoment of the two families in the same population /i.e. Maróti, 1980./ can exist only if the organisms start from a mirror symmetric initial arrangement and physical parameters later decide which of them will grow up to the right and which to the left family of enantiomorphy./Fig.5./

So mirror symmetry is the property of Fibonacci plants on family level in two aspects. First it is hidden in the Fibonacci numbered structure, second it is expressed in the two enantiomorphic families of Fibonacci plants.

1+1

				6	
				5	
				4	
				3	
6	5	4	3	2	1
				1	

1+2

				6	
				5	
				4	
11	9	7	5	3	1
10	8	6	4	2	
				1	

3+2

			16	14	12	
			13	11	9	
			10	8	6	
13	11	9	7	5	3	1
10	8	6	4	2		
			1			

3+5

			25	20	15	
			22	17	12	
			19	14	9	
31	26	21	16	11	6	1
28	23	18	13	8	3	
25	20	15	10	5		
22	17	12	7	2		
19	14	9	4			
			1			

8+5

			65	60	55	50	45	40	35	30	
			57	52	47	42	37	32	27	22	
			49	44	39	34	29	24	19	14	
56	51	46	41	36	31	26	21	16	11	6	1
48	43	38	33	28	23	18	13	8	3		
40	35	30	25	20	15	10	5				
32	27	22	17	12	7	2					
24	19	14	9	4							
			1								

Fig.4. Band pairs of normalized mosaic lattices of Fibonacci plant structures arranged according to the discrete development series. The series can be corresponded to the "discrete equation of motion" by part-band sliding operation with mirror symmetrical initial condition. This series belongs to the band pairs shown in Fig.3. which form one of the two enantiomorphic series /families/ to be shown in Fig.5.

This part of the family level model system represents the hidden conservation of mirror symmetry in the Fibonacci plant structures /Bérczi,1976/

8+13

			137	124	111	98	85	72	59	46	
			129	116	103	90	77	64	51	38	
			121	108	95	82	69	56	43	30	
			113	100	87	74	61	48	35	22	
144	131	118	105	92	79	66	53	40	27	14	1
136	123	110	97	84	71	58	45	32	19	6	
128	115	102	89	76	63	50	37	24	11		
120	107	94	81	68	55	42	29	16	3		
112	99	86	73	60	47	34	21	8			
104	91	78	65	52	39	26	13				
96	83	70	57	44	31	18	5				
88	75	62	49	36	23	10					
80	67	54	41	28	15	2					
72	59	46	33	20	7						
64	51	38	25	12							
56	43	30	17	4							
48	35	22	9								
			1								

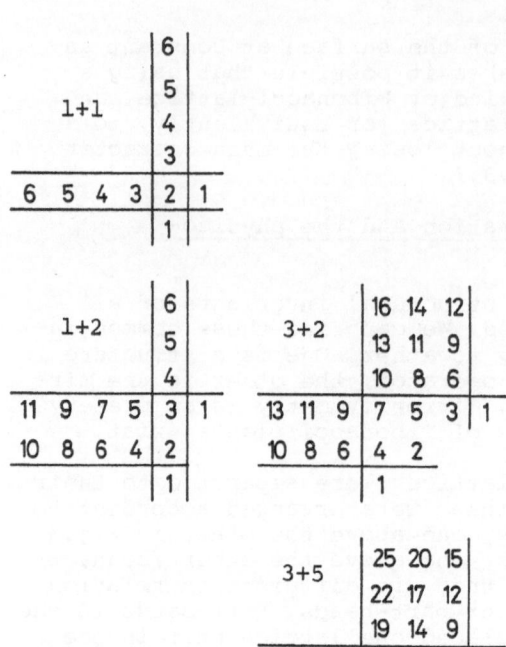

The sliding operation rearranges the lattice by sliding the bands one by the other in one lattice direction with one lattice unit distance. In the next step the operation slides bands in the other lattice direction so the operation works alternately changing lattice directions. This alternation of generator operation is the result of the decomposition of the motion according to lattice-directional components. This enantiomorphic pair of the two families of Fibonacci plant structures expresses the preservation of mirror symmetry which is inherent characteristic of the 1+1 band structure, the only common representative of the two enantiomorphic families.

The class level of deformation equivalent lattices on cylinder

We could recognize the effectivity of the part-band sliding operation both in connecting different Fibonacci structures and in deduction of all family members of related structures from the most simple, mirror symmetric 1+1 case. Now, let us use up this operation in extending the benefits of this mechanism to other, non-Fibonacci structures. Let us study the inverse of the deductive operation. The deductive part-band sliding operation builds up lattices with wider and wider bands on the cylinder, i.e. it works in sliding apart bands. But inversely, we can use the operation /always alternately changing its lattice directions/ in a sliding togethet fashion, too. This way we reduce the width of the bands. At the end of the sliding together operations a simplest structure of its family will appear. So this inverse operation, the sliding together operation can serve as reducing operation till the irreducible representative of the family of the starting case. This way to any kind of cylindrical normalized /to square mosaic lattice/ structures both the irreducible representative and its all family can be reconstructed. This way we have found a more general operation than which is useful only for Fibonacci structures. General classification of cylindrical normalized lattice mosaic structures can be given by the part-band sliding operation.

Before we carry out classification I show the reduction of the 7+4 structure. If we name one lattice element all its neighbours can be named according to the lattice positions in the band system. In the 7 unit wide band the cell numbers on the part bands follow each other as an arithmetical progression with 7 increment. Perpendicular to the earlier direction, in the 4 unit wide band the cell numbers on the part bands follow each ether as an arithmetical progression with 4 increment. In Fig.6. the named element is $\underline{a}$. The neighbours are named $\underline{1}$, $\underline{2}$, $\underline{3}$, ...etc. instead of $\underline{a+1}$, $\underline{a+2}$, $\underline{a+3}$, ...etc. because of the small place in cells. /Fig.6./

The sliding together operation begins in the direction perpendicular to the wider band. For the 7+4 structure the 2+1 will appear as irreducible representative of its family. The last step in this sliding together process is a virtual one in the sense, that in band winding to the left alone the tangent of by sliding /because of $\underline{a},\underline{3}$ neighbourhood changes to $\underline{a},\underline{2}$ one/ changes. The reducing sliding together operation results 2+1 structure as the irreducible representative for 7+4. Rule of alternating use of the sliding operations excludes further reduction.

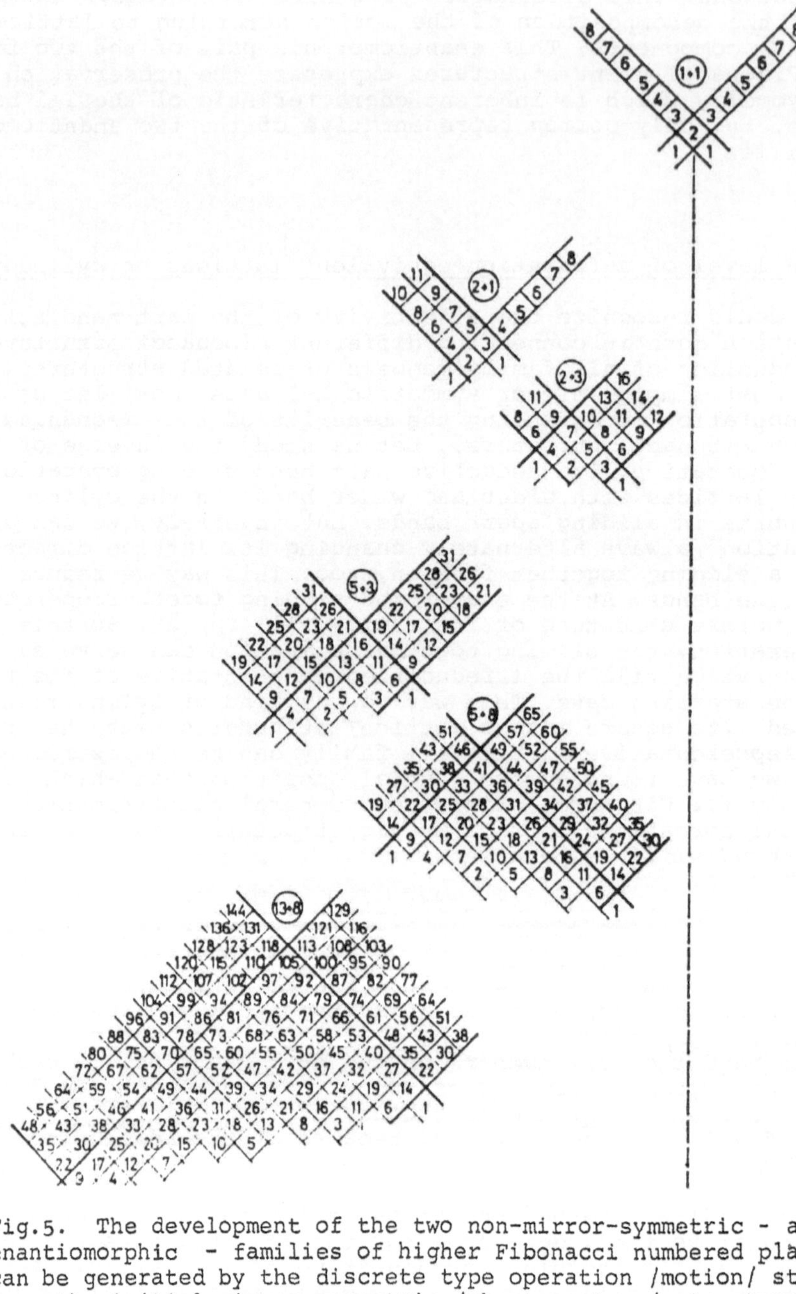

Fig.5. The development of the two non-mirror-symmetric - and so enantiomorphic - families of higher Fibonacci numbered plants can be generated by the discrete type operation /motion/ starting from the initial mirror symmetric /wheatear-type/ structure and violating its mirror symmtry at the first step of operation.

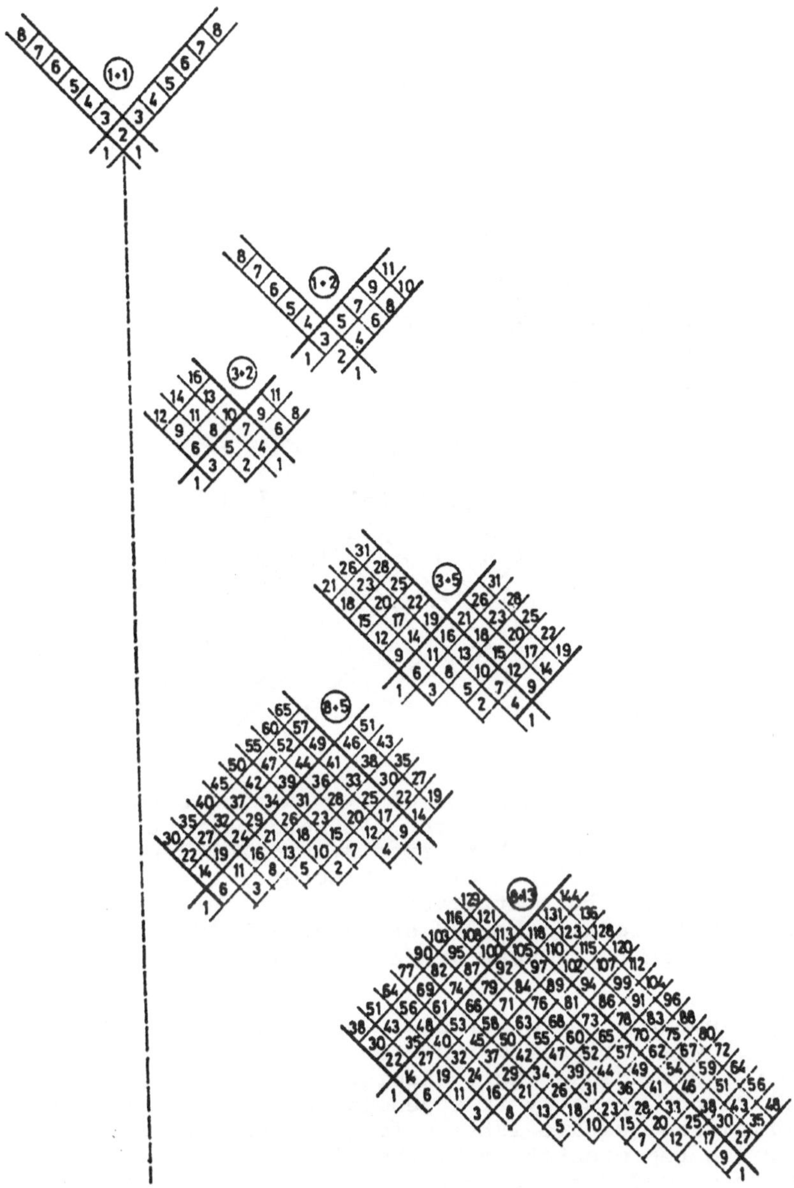
1•1
1•2
3•2
3•5
8•5
8•13

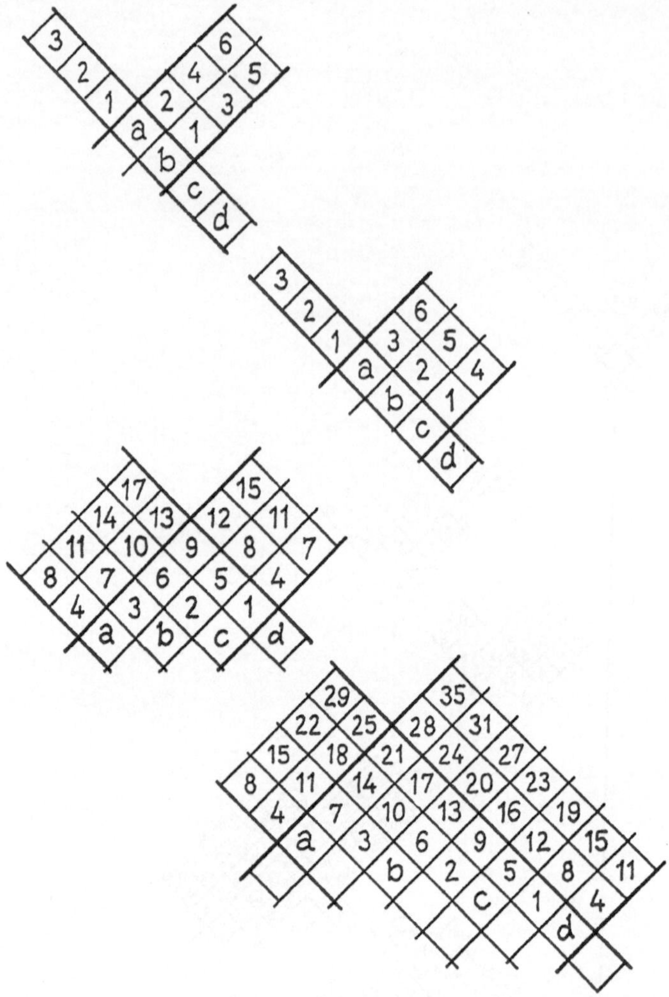

Fig.6. Steps in the reduction of 7+4 lattice structure to its irreducible representative, the 2+1 band-pair structure. The virtual sliding of unit band from 3+1 to 2+1 structure results similar lattice like that of the Fibonacci structure series, except, that from this structure there the sliding of 2 unit wide band will result the 2+3 structure, which does not occur in this Lucas numbered series.

In summary: the alternately used part-band sliding operation which was deciphered from the structure of Fibonacci plants helped us to conclude the following statements. A lattice given without mirror symmetry in a general position on the cylinder /in normalized, square lattice form/ 1. can be reduced to its irreducible representative by the sliding together operation and 2. can be extended to build up all of its family by both the sliding together and the sliding apart operations. By this operation disjunct families of deformation equivalent lattices can be given. All the families of deformation equivalent lattices occur in two enantiomorphic forms. These pairs of enantiomorphic families are separated /no common structure/ except the Fibonacci family where the 1+1 mirror symmetric structure is a common and so connecting structure between the two enantiomorphic sides.

What is the cause of the effective use of part-band sliding operation in classification of deformation equivalent families? The most important physiological cause is the growing up of Fibonacci plant organs. Growing up means continuous deformation of cylindrical lattice. But on rotational surfaces for lattices before and after deformation the following $r \sim \varphi$ relation must hold: $r \sim e^{\varphi}$. The discrete form of this exponential change of longitudinal lattice parameters /i.e.their numbers/ is visible in the changes of Fibonacci numbers in the structure - as a result of general crystallographic laws on the cylinder. On the other hand, the part-band sliding operation is a discrete two dimensional cylindrical representation of the generator addition operation of Fibonacci series on the number line. In other words: the Fibonacci series - with its generator process - is the projection of the part-band sliding operation /starting with the 1+1 structure/ products from the cylinder to the number line.

On the number line the generalized Fibonacci series have as building operation as the strictly named Fibonacci series has: by this process any two numbers may generate a discrete exponential number series by adding the two preceding numbers. Therefore any two numbers which characterize an irreducible cylindrical lattice representation can be corresponded to the the same two numbers on number line which build up their generalized Fibonacci series. Because of the correspondence the numbers of the generalized Fibonacci series will be the characteristic numbers of the cylindrical lattice family members. /In our example in Fig.6. the 2 and 1 generators of the Lucas series generate the 2, 1, 3, 4, 7, ... lattice-character numbers of the corresponded cylindrical mosaic structure family./

The cellular automaton model connects local and global structure

The cellular automaton model of Fibonacci plant symmetries was published elsewhere /Bérczi, 1985, 1987/. Therefore I refer to it in order to compare it to the global models /especially to the family level one/. On the orhet hand I should like refer to a local model which does not use up the role of golden angle in the structure description and deduction.

In cellular automaton model uniform cells arranged in a regular mosaic form a background. Initial conditions must be given. From this initial stage the cellular state changing operation develops the cell-arrangement. In our case the operation which was so useful in global structure description, must be transcribed to instructions for the cells. The preserved global character is reflected in the fact, that - on a cylindrical square lattice arrangement - instructions for cells are uniform /with the alternation between the two types of steps in the two main directions/. The neighbour changing instructions for cells are given in Fig.7. /cylindrical arrangement/.

The topological type local model and the earlier global ones emphasize the coherence of the local and global structure which is an inherent characteristic of crystallographic structures in general.

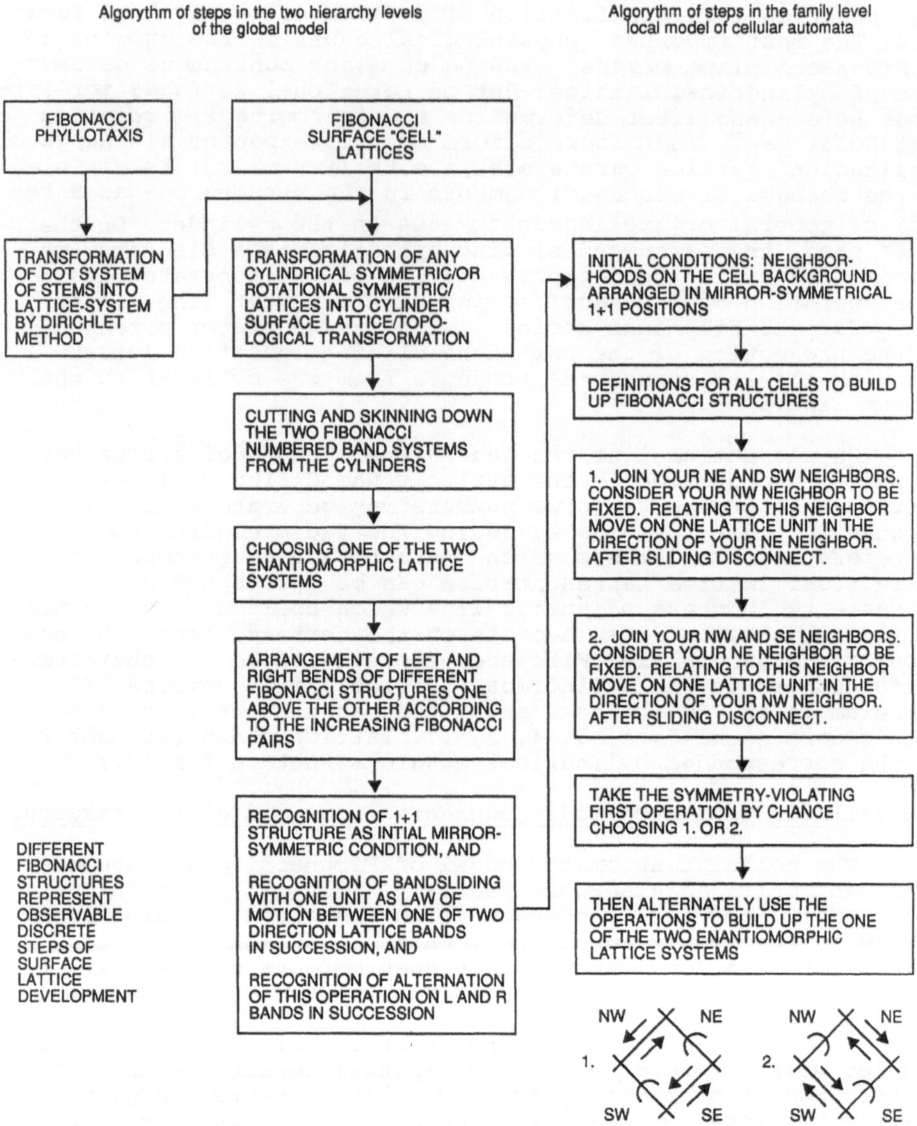

Fig.7. Comparison of a local and a global model of Fibonacci plant structures on family level./on the basis of Bérczi, 1987/ The algorythm of global model contains steps in model making from unit level, too /especially those transforming phyllotaxis to surface mosaics/. So this figure is a summary of our review considering the way of thinking from unit level to family level and both global and local variations of these models.

Summary

The symmetry principle appears in three different ways on the three hierarchy levels of Fibonacci plant structures. On unite level any Fibonacci plant can be characterized by a helical group, which can be divided - as a first approximation - to two Fibonacci number order subgroups. On family level enantiomorphic pairs of any kind of Fibonacci plants can be found. These pairs, however, can be connected into two enantiomorphic families by a deducing operation. These two families have only one common member: it is the mirror symmetric - for the operation the initial condition - structure of 1+1 band lattice. On the third level enantiomorphic families of Fibonacci type structures are the members of a classification which was carried out on the basis of general characteristics of lattice deformations on the cylinder. Deformation equivalent classes have irreducible representatives; every family has enantiomorphic pair.

The characteristics and consequences of the structure of Fibonacci plant symmetries on the three hierarchy levels were comprehended and reviewed by introducing a new operation. This global type part-band sliding operation was a useful tool in showing that Fibonacci structure was deeply connected to crystallographic regulations of lattices /or mosaics/ on the cylinder. On the other hand an interdisciplinary model making became possible. This model connected local and global type of models and topological, physical, botanical and crystallographyic structural descriptions of the phenomenon /without mentioning the structural role of golden angle between certain lattice positions/.

The new topological type operation made it possible to show the following:

1. Fibonacci structures conserve the initial mirror symmetry - although in a hidden form.
2. Biological surface structures of Fibonacci plants represent the general laws of deformations of cylindrical lattices during their development and evolutionary changes.
3. The existence and method of determination of irreducible lattice /mosaic/ structures on the cylinder.
4. Fibonacci plants form a deformation equivalent class among such type of lattice classes on the cylinder.

REFERENCES

Bérczi, Sz., 1976, Fiz. Szemle, 26:59.
Bérczi, Sz., 1978, Fizika 78. 87. p. Gondolat, Budapest
Bérczi, Sz., 1985, lect. pres. Conference on Intuitive Geometry 1985 Balatonszéplak, Hungary.
Bérczi, Sz., 1987, Organisational constraints on the dynamics of evolution, Manchester University Press, Manchester.
Bérczi, Sz., 1989, Deformációekvivalens rácsok osztályai a henge-

ren. /Classes of deformation-equivalent lattices on the cylinder/. Szimmetria és strukturaépités /Symmetry and structure-building/.J3-1441. Tankönyvkiadó, Budapest. /in press/.
Coxeter, H. S. M., 1961, Introduction to geometry. John Wiley and Sons, New York.
Erickson, R. O. 1983, in: The growth and functioning of leaves.53. Cambridge University Press, Cambridge.
Richter, P. H., and Schranner, R., 1978, Naturwissenschaften, 65: :319.
Weyl, H., 1952, Symmetry. Princeton University Press, Princeton.
Wigner, E., 1964, Proc. Natl. Acad. Sci. U.S.A. 51:956.

STRUCTURES, FORMS, PATTERNS AND PERCEPTION: A BRAIN-THEORETIC POINT OF VIEW

P. Erdi

Central Research Institute for Physics
Hungarian Academy of Sciences
H-1525 Budapest, P.O.B. 49, Hungary

1. INTRODUCTION

The very essence of the neural, separating it from all other parts and aspects of the living system, is its unbelievably complexity. The brain can be considered, as a prototype of the hierarchical structures. From methodological point of view it can be investigated at molecular, cellular, synaptic, network and system level.

Experimental neuroscience, i.e. the modern neuroanatomy combined with biochemical and microphysiological methods, in one side, serves many information about the structural details of the geometry of nerve cells and its processes, in other side, forms the substrate of theoretical studies about the organization principles of the higher neural centers.

Before starting to analyze neural systems, some general remarks on the physical structures have to be made. Physical structures can be categorized into two classes: **static** and **dynamic** structures appear as distinct forms of the world. Static structures are maintained by the large interacting forces among their constituents. The perturbation of the environment causes very slight effects only, at least in the stability range of the (structurally stable) structures. Applying a more intensive perturbation the structure breaks down completely. The new, 'structureless' structure is also static, more precisely '... the final situation may be topologically very complicated, but generally it behaves like a static form...' (Thom 1975, pp. 101). Crystals are typical examples of the static structures, and they are subjects of traditional symmetry analysis. The dynamic structures are maintained by the interaction between the system and the environment. The system itself, is in permanent material, energetic and informational interaction with the external world, thus it can be qualified as an open system. Dynamic structures, as regular spatial patterns of

Symmetries in Science IV
Edited by B. Gruber and J. H. Yopp
Plenum Press, New York, 1990

biological systems are the results of symmetry-breaking morphogenetic mechanisms.

Neural structures, though they are constituted by chemical molecules, can not be ultimately analyzed in terms of chemistry. Their order and symmetry properties perceived (and reviewed here) by a brain - theoretician are different from symmetries seen '... through the eyes of a chemist', (Hargittai & Hargittai, 1987), even he had got his formal training in chemistry.

An attempt is given here to review some properties of neural structures and forms related to their functions in terms of symmetry, asymmetry and symmetry-breaking. First, 'elementary' neural structures, i.e. molecules, neurons and their dendritic patterns and synapses are studied (Section 2). It seems to be evident that the nervous system is composed of 'building blocks' of regular structures. The 'modular architectonic' concept, as a general anatomical organizing principle of the nervous system, (and of the cerebral cortex, in particular) has been rather extensively discussed recently (Szentágothai 1983, 1987). Some aspects of the cortical architectonics will be emphasized here (Section 3). The problem of the representation of the sensory information in the cortex, and some concepts on perception are discussed in Section 4. Viewing the brain as a mixture of symmetric and asymmetric devices, we can not avoid of mentioning the cerebral asymmetry and its relationship to aesthetics (Section 5). Conceptual and mathematical models of pattern formation are shortly studied (Section 6). We conclude, that the emergence of the marvellously complex structure of the nervous systems is the result of the interplay between incredible precise genetically determined wiring and self-organizing mechanisms utilizing the beneficial role of environmental noise (Section 7.)

2. ELEMENTARY STRUCTURES

2.1. On the role of the molecular symmetry:

One remark

The most important form of the communication between neurons is the neurochemical transmission through synapses. Neurotransmitter molecules released by the presynaptic cell react with receptor molecules bound at the postsynaptic membrane surface. Pharmacological substances can, competitively to transmitter molecules, bind to receptors. It was demonstrated (Pert & Snyder 1973) that between the two mirror-symmetrical molecular forms of a morphine derivative only the levorphanol binds to morphine receptors and stops pain, while the dextorphan does not react specifically. (Fig.1.) The phenomenon can be explained by some key - lock mechanism. (Further researches led to the discovery of 'endogenous morphins', endorphins, and of enkephalins, which proved to be internal pain reducing peptides.)

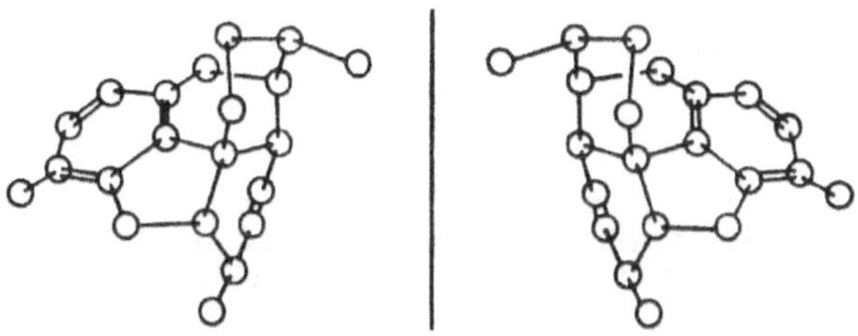

Figure 1. Mirror-symmetrical forms of a morphine derivative. Only one of the two optical isomers, namely levorphanol, binds to morphine receptors and stops pains.

2.2. On the 'polarity' of neurons

It is well known already from the classical histological studies, that neurons exist in rather diverse forms. The basic parts of a nerve cell are the soma, and the processes that project from the cell. In some primitive neurons the processes (i.e. the axons, and the dendrites) are indistinguishable. These neurons are called **isopolar**, the name of the complementary class is **heteropolar**. According to the number of process emerging from the soma nerve cells can be classifed as **unipolar**, **bipolar**, and **multipolar**. Some types of neurons, distinguished on the basis of number and differentiation of processes are shown in the Figure 2.

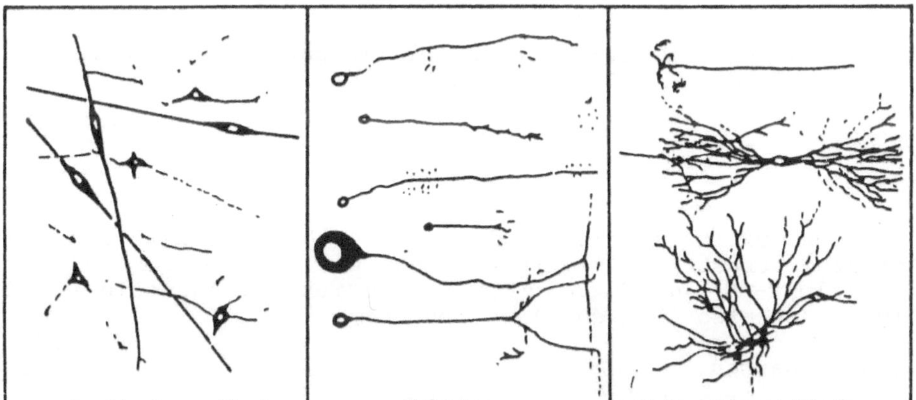

Figure 2. Polarity of neurons (after Bullock and Murridge 1965)

Newer studies (Black & Baas 1989 suggest that the intrinsic polarity of neurons might be derived from the asymmetry of the tubulin molecule.

2.3. Dendritic patterns

Though dendrites were originally defined as processes specialized to be the receptive part of the neuron, it turned out that sometimes they can be presynaptic elements. The dendritic arborization can be classified based on (1) 'randomness' versus 'regularity' in regard to branching of the various dendritic segments resulting in a continuous spectrum ranging from the 'radiate' to the 'tufted' types of arborization, and (2) the degree of deviation of the individual branches from a radiated and rectilinear course. (Szentágothai & Arbib 1974). The main types of patterns were illustrated by Ramon-Moliner & Nauta, 1966), see Fig. 3.

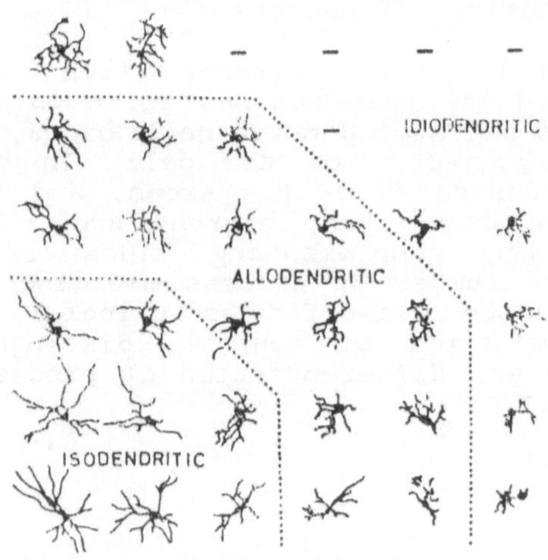

Figure 3. Varieties of dendritic patterns found in the brainstem of the cat.

Dendritic patterns determine the organization of ordered structures. Figure 4. shows the regular distribution of 'alpha-ON' ganglion cells in the retina. (Waessle et al 1981, see also Changeux 1985, pp. 64).

2.4. Synapses

A synapse is the site of contact at which a presynaptic neuron can transmit a signal to the postsynaptic cell. While the majority of the synapses is axo-dendritic, others, as axo-somatic, dendro-dendritic, somato-somatic, axo-axonic, also may occur. (Figure 5, Shepherd 1983, pp.80).

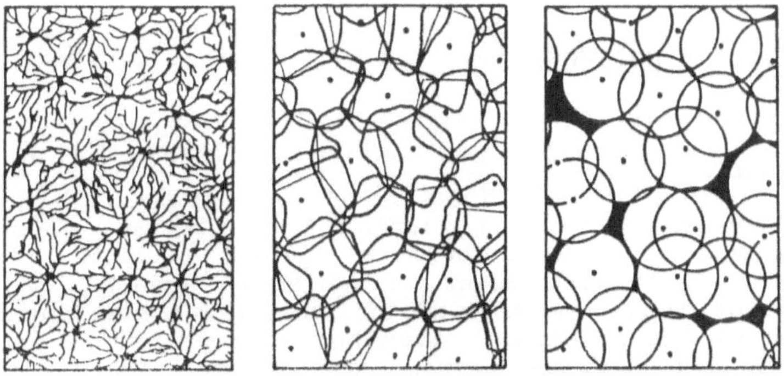

Figure 4. Regular distribution "alpha-ON" ganglion cells in the retina. In the center diagram, dendritic branching is represented by continuous lines, while in the right side of the figure circles of constant radius have been drawn around the point line center.

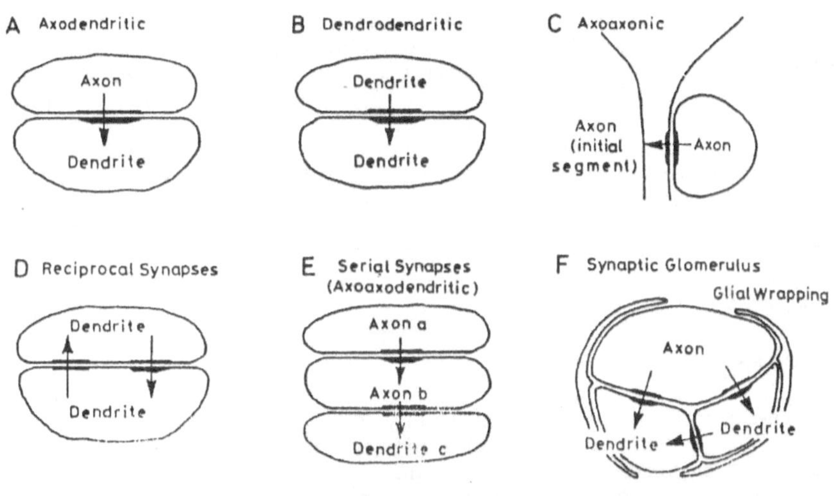

Figure 5. Types of synaptic arrangements.

Further comments, e.g explanations for the functional consequences of the anatomical diversity, are beyond our scope. For a recent review on 'The Nerve Cell and Its Processes' see Killackey (1985).

3. CORTICAL ARCHITECTONICS

3.1. The modular architectonics principle:

some remarks

Experimental facts resulting from anatomy, physiology, embriology, and psychophysics give evidence in the vertebrate nervous system of highly ordered structure composed of 'building blocks' of repetitive structures. The building blocks or modular architectonic principle is rather common in the nervous system. The modular architecture of the spinal cord, of the brain stem reticular formation, of the hypothalamus, of the subcortical sensory relay nuclei, of the cerebellar and mostly of the cerebral cortex has been reviewed (Szentágothai 1983, 1987). The cortex can be regarded as a two-dimensional sheet, a few millimeters in thickness, stratified into six parallel horizontal layer. Vertical, columnar organization of neurons having similar receptive fields was observed in the somatosensory cortex (Monutcastle 1957) and in the visual cortex (Hubel & Wiesel 1959) by physiological methods. After the anatomical demonstration of the so-called cortico-cortical columns (Goldman & Nauta 1977) it was suggested (Szentágothai 1978) that the cerebral cortex be considered on large scale as a mosaic of vertical columns interconnected according to a pattern strictly specific to the species. Having been motivated by the pioneering work of Katchalsky (see Katchalsky et al 1974) on the dynamic patterns of neural assemblies Szentágothai offered to interpret the cortical order in terms of 'dynamic structures' instead of applying some 'crystal-like' approach. The more precisely a system is specified, the greater is the danger of mistakes. Spatially ordered neural structures are the product of some **self-organizing** mechanism. Even 'the essence of neural' may be its self-organizing character (Szentágothai & Erdi, in press).

3.2. Ontogeny

The problem of the ontogeny of organs and of organisms is strongly related to the dichotomy between the strict genetic determinism and the environmental controlled activity-dependent mechanisms. The **radial unit hypothesis** (e.g. Rakic, 1988) gives a framework to find the proper role of both mechanisms in forming the cortical structures and patterns. According to the main points of the hypothesis, the cortical neurons generated prenatally (i) form a pre-ordered proliferative, ventrcular unit, (ii) migrate along radial glial guides, (iii) during neural migration the spatio-temporal order is preserved. (Fig. 6.)

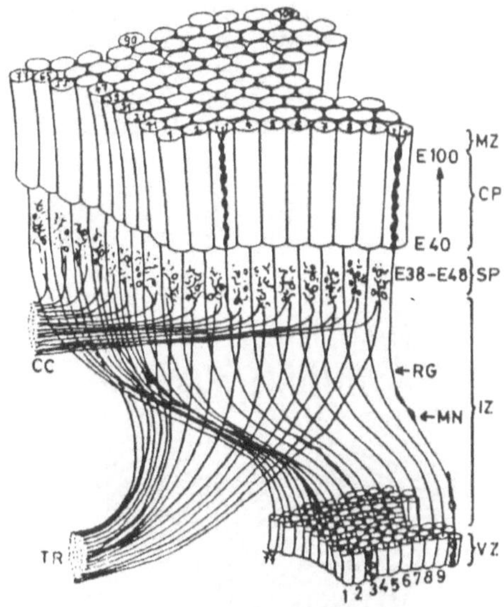

Figure 6. Spatial order of the proliferative, ventricular zone (VZ) is preserved within the cortical plate (CP) during cerebral development.

The radial unit hypothesis is in accordance with the selective stabilization hypothesis (Changeux & Danchin, 1976). This latter offers a third option between the 'preformist and the empirist' hypothesis. The first states theat neuronal connectivity is pre-specified genetically, the second (over)emphasizes the role of the activity of the system to specify its connections. The main conceptual advantage of the selective stabilization hypothesis is that it offers a gene-saving mechanism for specifying ordered neural structures.

3.3. Ocular dominance columns

The visual cortex is considered as the best paradigm of modular organization of the neural centers. Ocular dominance columns or stripes are characteristic examples of such kinds of organiztion. In normal cats and monkeys the afferents from lateral geniculate nucleus (LGN) laminae corresponding to the two eyes innervate common target structures, e.g. layer IV of the visual cortex and form partially overlapping bands. This alternating termination of LGN afferents showing about 300-400 um wide periodicity is thought to be the anatomical substrate for the physiologically defined ocular domains (Hubel & Wiesel 1972, Hubel, 1982, Wiesel 1982; some further readings: LeVay et al 1975, Ferster & LeVay 1978, Schatz et al 1977, Swindale 1982, Mower et al 1985, Anderson et al 1988).

It is likely that ocularity domains are incompletely formed in monkeys and absent in kittens immediately after birth. While the normal development leads to near-periodic spatial patterns, visual deprivation during the critical period results in severe symmetry-breaking of the width of ocularity domains (Fig. 7).

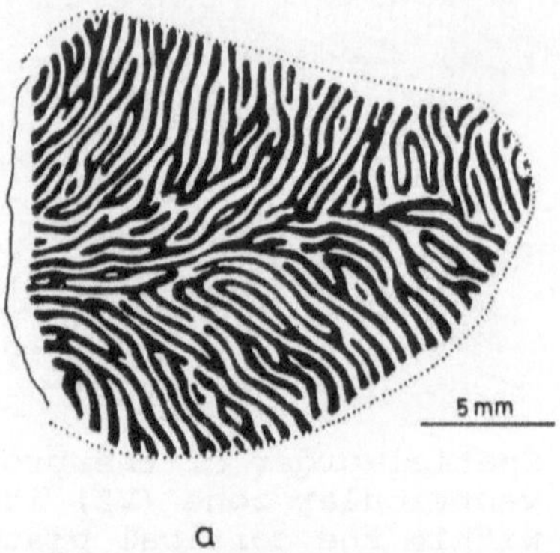

Figure 7a. Reconstruction of ocular dominance columns (layer IVC) of the visual cortex of a normally developed macaque monkey. (From Hubel, D.H. and Wiesel, T.N., 1977).

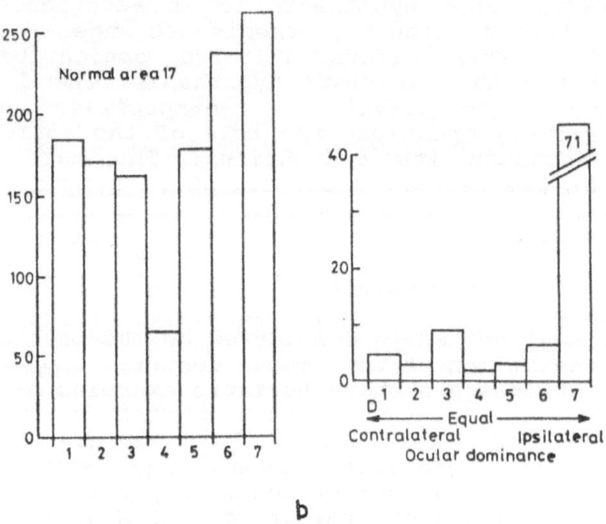

Figure 7b. Ocular dominance histograms in normal and monocularly deprived versus macaque monkeys. Monocular deprivation caused a significant symmetry breaking in the ocular dominance.

Activity patterns heavily influence the formation of ocularity domains. Treatment with tetrodotoxin of one eye of a kitten from 2 to 6 weeks of age prevent the normal development of ocular dominance columns. It was demonstrated that **spontaneous** activity (having random character) must be sufficient (Stryker & Harris 1986; for a general review of activity-dependent self-organizing neural mechanisms, see Singer 1987).

3.4. Orientation columns

It is well known neurophysiological fact that neurons in the primary visual cortex give selective reponse for the orientation of the stimulus. (Hubel & Wiesel 1974). More precisely, (i) the change of orientation specificity with the movement of the recording electrode through the cortex is 'regular'; (ii) the change of orientation with the movement of electrode is clockwise for part of the trajectory, counterclockwise for another part; (iii) a sudden jump might ocur in an otherwise smooth progression of the orientation along a straight line in the cortex.

The **anatomical substrate** of the orientation columns is much less known than that of the ocular dominance columns. The application of the modern anatomical techniques, i.e.

(i) the 2-deoxyglucose, a metabolic marker for labeling active neurons (Hubel et al 1978); (ii) voltage-sensitive dyes (Blasdel & Salama 1986) and mostly studies of the distribution of the mitochondrial enzyme, cytochrome oxidase (Wong-Riley 1979, Horton & Hubel 1981) gradually offered newer and newer ideas not only on the functional structure of the orientation columns, but on the whole visual cortex.

The Hubel - Wiesel experiments reinforced the hypothesis on the columnar arrangements of cells having similar orientation selectivity, but instead of being pillar, the columns seemed to be parallel slabs that intersect the surface.

A centric arrangement of orientations has been suggested by Braitenberg & Braitenberg (1979). **Circular symmetry** around centers, either radially (Fig 8a), or along concentric circles (Fig. 8b) was suggested as natural explanation of the facts observed. An assembly of neighbouring columns of the assumed basic modules would be interpreted as 'hypercolumns' of 2-3 mm diameter. Two types of hypercolumns (**d**extrorotatory and **l**evorotatory) occur in alternating sequence (Fig. 9.), according to theoretical analysis (Braitenberg 1986, see also Goetz 1987, 1988).

3.5. The superposition of the two columnar systems

further difficulties

Albeit previously we used the term 'hypercolumns', as an organization concept of the orientation columns, it was introduced by Hubel & Wiesel (1974), as a conceptual device

Figure 8. Alternative ways of interpretation of centric arrays of orientations (a) radial (b) concentric. Arrows: movement of an electrode through the cortex produces regular changes of orientation.

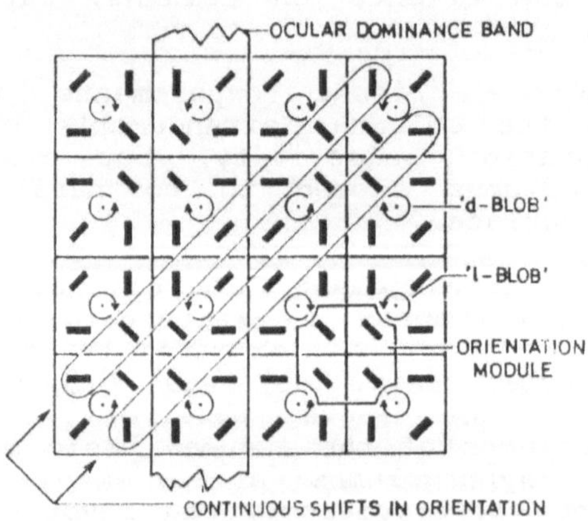

Figure 9. Illustration of the two types of hyper-columns having different "handedness" ("l-type" and "d-type"). For the meaning of blobs see Ligingstone and Hubel, 1982.

for visualizing the idealized relationship between ocular dominance columns (slabs) and orientation columns (slabs). According to this approach, a hypercolumn consists of a pair of ocular dominance columns and a complete set of oreintation columns (Fig. 10a). Of course, this illustration is not in accordance with the Braitenberg's view of circular symmetry of orientation.

The most important finding, which has modified this 'ice-cube' model of the primary visual cortex was the demonstration of citochrome oxidase-rich 'blobs' (Livingstone & Hubel 1981). It was shocking, that blob cells showing high spontaneous activity and selectivity e.g for color and brightness, did not exhibit orientation selectivity. A review (Martin 1988) on the role of cytochrome oxidase for our view on visual perception is highly recommanded).

We can not go here into the details of the physiological properties of the 'blobs' and 'interblobs', not only because it beyonds our scope, but also because of the lack of clarity of concepts, at least for the time being. A picture of the modified 'ice-cube' model is given here (Fig. 10b, based on Martin, 1988). However, this modified picture reinforces the view that the visual system is a functionally parallelly organized, distributed system, where form, colour and spatial information are processes along independent channels (Livingstone 1988, Zeki & Shipp 1988).

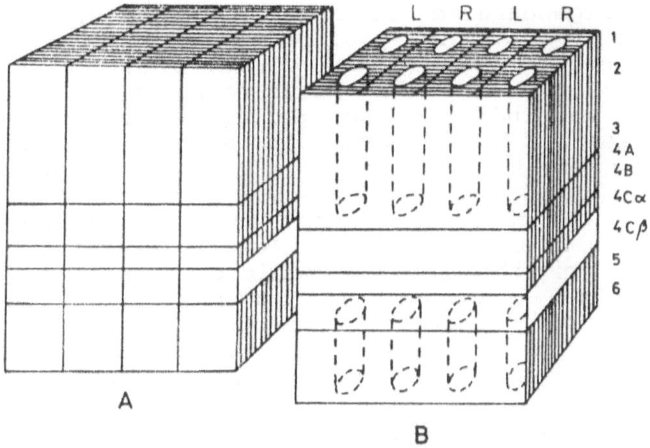

Figure 10. The functional architecture of the primary visual cortex. (A) the original 'ice cube' model illustrating the connections among ocular dominance columns and orientation columns. (B) the modified 'ice cube' model takes into account the existence of the 'non-orientation selective' blobs.

3.6. Cortical plasticity

Making our analysis from the point of view of symmetry, what it is mentioned first is, that asymmetries in the activation of afferents from the two eyes imply corresponding asymmetries in the geometry of ocular dominance columns. The modifiability of the width of columns, however, is restricted for the so-called critical period of the normal development. Mechanism of plasticity of ocular dominance columns is activity-dependent.

The columnar organization in the domain of orientation selectivity shows also experience-dependence and plasticity. Asymmetries in the visual environment causes asymmetries in the orientation preferences of the cortical neurons. Orientaion columns also can give plastic responses for external influences, as monocular deprivation. Perhaps the mechanisms of plasticity of orientation columns are less clarified, than one of the ocular dominance columns.

(For a review of activity-dependent self-organizing mechanisms of cortical plasticty, see Singer 1987.)

4. REPRESENTATIONS

4.1. Cortical representation of sensory surfaces

Sensory information is represented in the cortex according to topic principles, i.e. in a topographically ordered way. Skin surface and retina are mapped into the cortex somatotopically and retinotopically, respectively. Topographic principles preserve tha spatial relationship, i.e. the neighbourhood relations at large, but local variations may occur at the level of individual cells. Though retinotopic maps of the cortex might be considered point-to-patch rather than point-to-point relationships, topographic mappings to be described by mathematical functions assign a location $y \in \mathbf{R}^2$ in the target area to every point $x \in \mathbf{R}^2$ in the source area.

It was suggested (Schwartz 1977), that topographic mappings should be **conformal** ('angle preserving'), and particularly for the retinogeniculocortical case, complex logarithm. This hypothesis is based on empirical facts referring to the geometrical structure of the retina and cortex: the retina shows evidence of radial symmetry, and cortical structures exhibit translation symmetries. Complex logarithm fulfils the criteria of mapping radial symmetries into translational symmetries.

The determination of the functional form of the topographic mappings in the visual system is a paradigmatic example of demonstrating the interplay between experimental neurobiology (i.e. neuroanatomy and neurophysiology) and mathematical analysis (i.e. theory of two-dimensional mappings, complex analysis) and led to the emergence of the 'computational anatomy' (Schwartz 1980).

The complex logarithm mapping has a well known invariance properties under the size and rotation transformations, as it is illustrated by Figure 11.

In spite of the appealing properties of the complex logarithm mapping there is no theoretical reason to assume its general validity. In fact, complex power functions have been suggested for describing the overall behaviour of the cat's visual cortex (Figure 12), (see Mallot 1985, Seelen et al 1987).

Since perceptual constancy, i.e. invariant perception under certain kinds of transformations, is a prerequisite of having a proper retinotopic mapping, group theoretic analysis is needed for studying the eventual existence of more general invariances in pattern recognition. A rather general analysis has been given in terms of Lie algebra already in 1966 (Hoffman 1966).

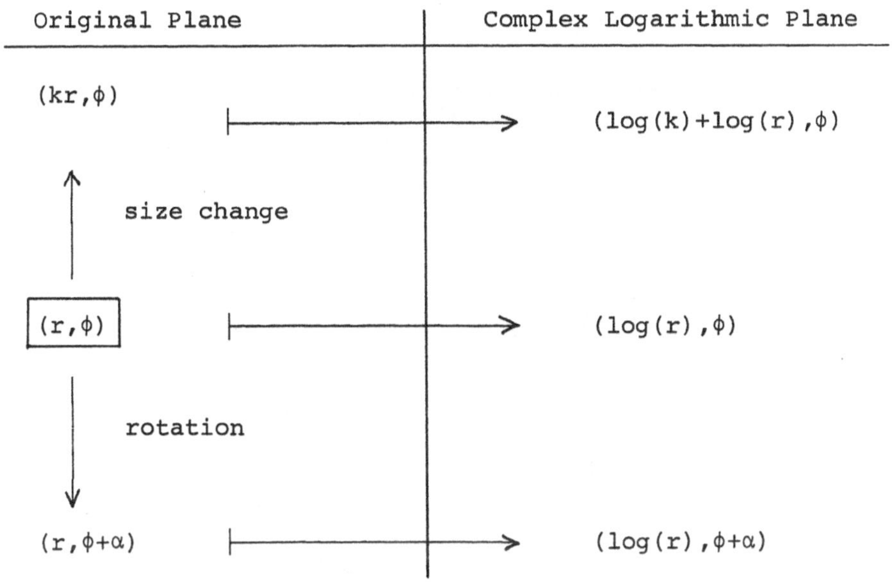

Figure 11. Size change and rotation in the image plane implies shift in the complex logarithmic plane

4.2. Perception of symmetries

It is a rather well-known empirical fact that symmetry influences the perception of form. The perceptual value of symmetry has been recently emphasized (Locher & Nodiner 1989).

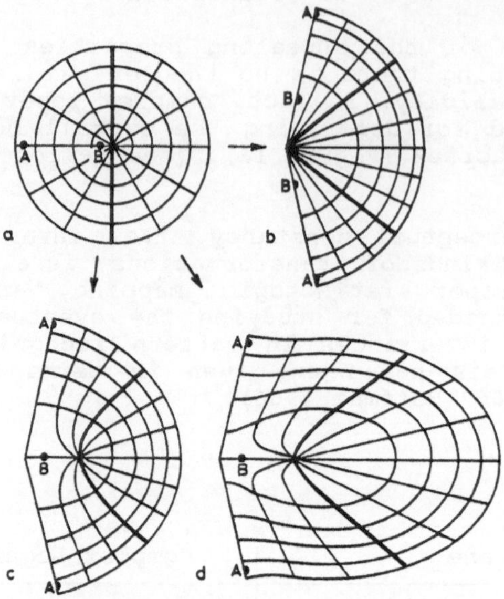

Figure 12. a) polar grid; b) transformation of a polar grid by a complex power function; c) by an eccentric power function; d) by a two-step modification.

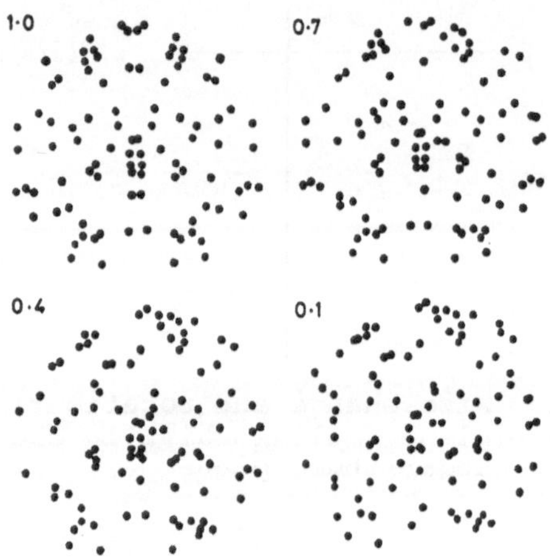

Figure 13. Degradation of symmetry by substituting paired dots with randomly placed dots. Proportion of "paired" and "random" points are shown by the numbers.

The sensitivity and versatility of mirror symmetry detection has been demonstrated by psychophysical methods (Barlow & Reeves 1979). One of their experiments (Figure 13) gives evidence for the continuous (in contrast to the all-or-none) character of the symmetry property. Each pattern containing 100 dots is the mixture of symmetrically paired and of randomly located points. Even the proportion of the paired dots is 0.4, symmetry is rather easily detected. The underlying neural mechanisms of symmetry detection might be connected to the problem of spatial vision (see 4.3).

The search for efficient symmetry finding algorithms is one of the most challenging problems of 'computational morphology'. A survey of algorithms working in two and three dimensions has been give (Eades 1988). The time complexity of the overwhelming majority of the algorithms is in the order of n log n, where n is the finite number of points or line segments of the set to be analyzed.

4.3. Spatial visual perception: psychophysical versus neurophysiological approach

Psychophysical evidences suggest that visual information is processed in parallel by separate channels, tuned to different **spatial frequencies** (see the seminal paper of Campbell & Robson 1969). According to this point of view the visual cortex can be described as a kind of Fourier analyzer, while neurophysiological studies suggest (e.g. Hubel & Wiesel 1962, Barlow 1969) that the visual cortex might be interpreted as a population of feature detectors.

A lot of works have been made in order to unify the results obtained by the two approaches. Some physiological and anatomical evidence supported the multiple spatial channel view. There are some substructures of the visual cortex that selectively respond to limited ranges of frequency (Hubel & Wiesel 1968, Maffei 1978), a 2-deoxyglucose study had suggested a columnar organization of spatial frequency specificity in primary visual cortex (Tootell et al 1981).

The connection between spatial frequency information and aesthetics value has been demonstrated by Rentschler et al 1988). They showed that '...styles of painting from the Middle Ages to neoimpressionism can be characterized by the preservation of the structure of physical images rendered through different types of spatial filters...'.

4.4. Symmetries of impossible forms

An impossible form presented here (Figure 14) is the work of the Hungarian visual artist Tamás Farkas (Farkas & Erdi 1985). The term 'impossible' refers to the fact that a structure can not be realized in three dimensions. Impossible forms, however, may be interpreted by certain mental procedures.

Figure 14. Continuous form organized in eight directions. Considering the figure either as a "planar" or as a spatial "impossible" form you might find different symmetry elements.

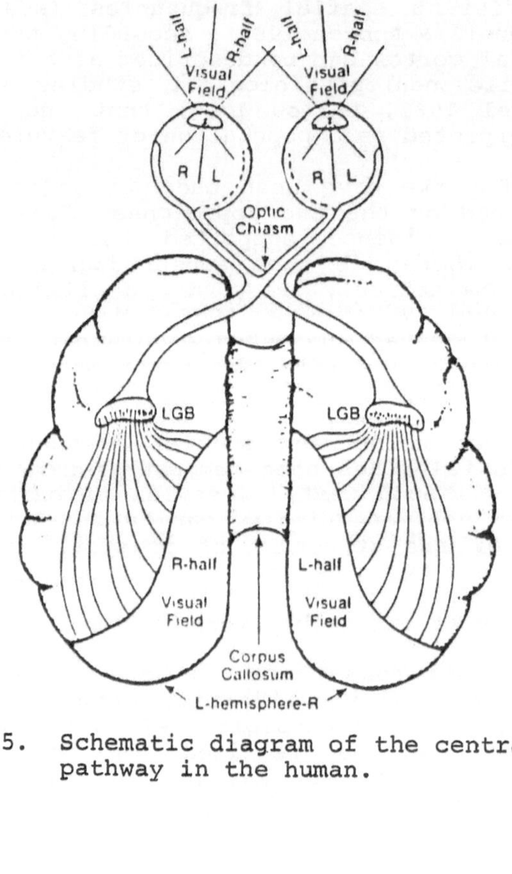

Figure 15. Schematic diagram of the central visual pathway in the human.

The unusal appearance of this figure results from the obvious cônflict between the planar and the quasi-planar form of the plane on one hand, and the unification of different views on the other.

Even for 'possible' forms is true that the original spatial forms and their projections differ in symmetry elements. (Let you think for the 2D representation of the cube). Farkas's figure can be analyzed from the point of view of its symmetry, i.e. you might ask whether what are the symmetry elements of the planar and of the 'spatial' form?

5. CEREBRAL ASYMMETRY AND AESTHETICS

Albeit the brain is a single system, it is composed of two halves (hemispheres) linked by bundles of nerve fibers. One of the characteristic properties of the brain organization is the decussation of the nervous pathways. Figure 15. is a schematic diagram of the central visual pathways in the human illustrating the partial decussation of the optic tracts.

In spite of its physical symmetry, the brain is functionally asymmetric. Handedness, i.e the difference in the abilities of the two hands is one reflection of this asymmetry.

In particular, many data on functional lateralization came from observations of brain-damaged patients. Current interests in the left brain - right brain problem has been tractable by making split-brain researches. Split-brain patients have been undergone surgery to cut the main cortical pathways (i.e. the corpus callosum) that normally connetcs the cerebral hemispheres.

Many problems on cerbral assyemtry related to aesthetics have been studied, language function in the hemispheres, the problem of handedness, the phylogenetic and ontogenetic development of asymmetry, just to mention a few of them. It is well known, that Roger Sperry (see e.g. Sperry 1974) shared the Nobel prize with Hubel and Wiesel in 1981 for his studies on split-brain patients; for a review of the left brain - right brain problem see e.g Springer & Deutsch 1981.

Cerebral asymmetry might have a significant role in creating and accepting artworks. The interaction between the 'spatial' right hemisphere and the 'temporal' left hemisphere is particularly important, since each half of the brain is selectively activated. It was suggested (Levy 1988) that '...artists, whether musicians or painters, have an unusual degree of bilateral repesentation of artistic skill, which permits superior interhemispheric integration. Experience in the artistic endeavor further promotes a concordance of hemispheric representation and structures and collaboration between the two hemispheres.'

6. PATTERN FORMATION

6.1. Biological pattern formation and neuronal forms

The term 'pattern' is one of the most popular expression of present-day science. Biological pattern generating mechanisms, occuring at different hierarchical levels, might have quite common structure (see Rosen 1981).

One main problem of developmental biology is whether '...how do cells in different positions become programmed to follow different developmental pathways and thus to produce the chracteristic anatomy of the organism?...' (Slack 1987). Answers are associated to the concepts of positional information (Wolpert 1969, Tsonis 1987). According to the assumption of that theory positional information is provided by morphogenetic gradients. Therefore the use of reaction - diffusion models, applied already earlier as models of biological pattern formation, seemed to be relevant.

A rather general class of the equations applied are:

$$\frac{\partial c_i(x,t)}{\partial t} = f_i(\underline{c}(x,t)) + \mathcal{D}_i \tag{1}$$

where $c_i(x,t)$ is the concentration of the ith component, the function f prescribes the kinetcis due to the chemical reactions, and $\mathcal{D}_i$ is the redistribution operator containg term due to diffusion, convection etc.

The Turing - Gierer - Meinhardt (TGM) model (Turing 1952, Gierer & Meinhardt 1972) is a specific example of equation (1) to describe the activator (a) - inhibitor (b) interactions:

$$\frac{\partial a(x,t)}{\partial t} = f(a(x,t),b(x,t)) + D_a \Delta a(x,t) \tag{2}$$

$$\frac{\partial b(x,t)}{\partial t} = g(a(x,t),b(x,t)) + D_b \Delta b(x,t)$$

A three-variables extension of the TGM model has been applied to describe network formation. The model led to uniform behaviour (Figure 16). For a review on the relevance of reaction - diffusion theory see Harrison 1987.

It was argued (Schierwagen, in press) that in accordance with the general view, that the TGM model explains the spatial distribution of the morphogens only, and not the shape of the anatomical structure, dendritic pattern formation of neurons can better described by other models. Since it was suggested, that neural dendritic branching might have a **self-similar** character, its development can be modeled by fractal generating growth models.

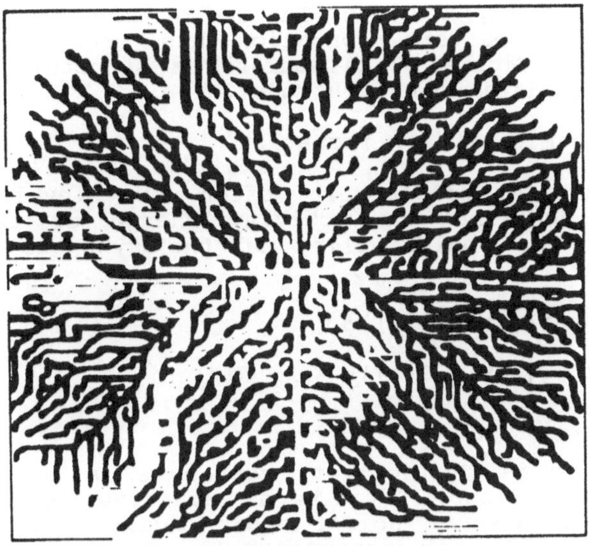

Figure 16. An uniform branching pattern, generated by the Turing-Gierer-Meinhardt model.

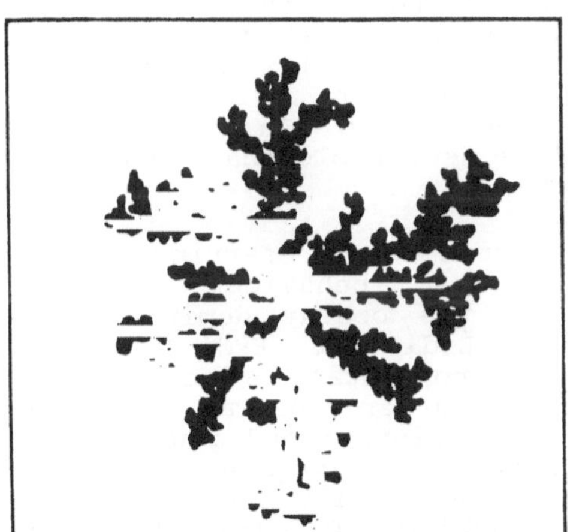

Figure 17. A fractial branching pattern, generated by limited growth.

Some types of growth models, often associated to fractal generation, have been established and studied in the last couple of years (see e.g. Stanley and Ostrowsky 1986, Peliti & Vulpiani 1988). Some of the most often investigated models are the Eden model, diffusion limited aggregarion, diffusion limited growth etc. The use of such kind of models for biological pattern formation has been demonstrated (Meakin 1986). Figure 17 illustrates a structure generated by a diffusion limited growth model.

6.2. Neurodynamic pattern formation

Neurodynamic models are often two levels dynamic systems to describe the activity change of the single neurons and the modifiability of the synaptic efficacy. Schematically:

$$\Delta c(t) = f(c(t), S(t))\Delta t$$
$$\Delta S_{ij}(t)=g(c(t), S(t))u\Delta t + R(t) \qquad (3)$$

Here c(t) is the activity vector, S is the matrix of the synaptic efficacy R(t) is an additive noise term to simulate environmental noise, and u scales the time. The specific form of f is determined by the velocity of the transfer of the presynaptic information, and the spontaneous activity decay. The function g contains the modification due to local cooperation among neighbouring synapses, and global competition of ingrowing axons for receptor molecules.

It is accaptable - at least as a working hypothesis - that ontogenetic neural development, plasticity, memory and learning can be associated to the modifiability of synaptic efficacies. In accordance with the spirit of Hebb's cell assembly approach (Hebb 1949), synaptic efficacies change due to correlation between presynaptic and postsynaptic elements. 'Anti-Hebb' rules (Palm 1982, Barlow & Foldiak 1989) may have some neurobiological relevance in describing decorrealation effects.

Simulation experiments referring to the ontogenetic formation and plastic behaviour of the retinotectal connections and of the ocular dominance columns (Érdi & Barna 1984, Barna & Érdi 1986, Érdi & Barna 1987) gave the evidence, that environmental noise is indispensable in establishing globally ordered neural structures. From formal point of view the functions f and g were specified in order to set up the model. The concept of environmental noise during growth has succesfully been adopted in biological context (May 1972, Nobile & Riccardi 1984). Furthermore, the theory of 'noise-induced transition' (Horsthemke & Lefever 1984) states that noise might play active and constructive role in the organization of ordered structures. Noise superimposed might change the qualitative behaviour of the deterministic system by destabilizing temporary, metastable structures. Figure 18 (Érdi & Barna 1984) visualizes the values of the matrix S of the formed retinotectal connections simulated by an (over)simplified model. While

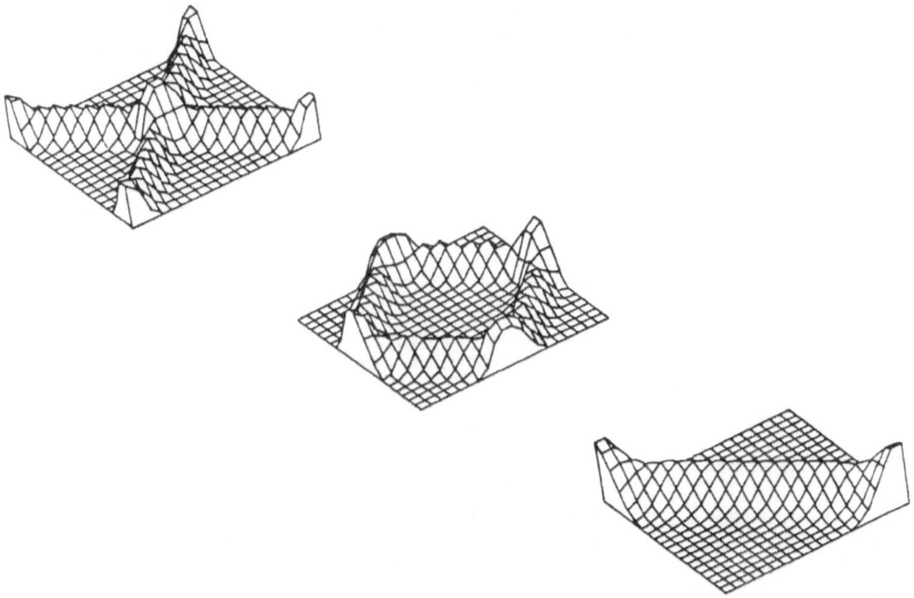

Figure 18. Visualization of the qualitative behaviour of a self-organizing algorithm. Values of the synaptic strength are plotted. "Double-diagonal" and "crater-like" solutions are produced by deterministic learning rules. The "diagonal" attractor associated to topographic order emerged by including a small additive random term.

the first and second part of this figure illustrates attractors established by a deterministic learning rule, the third part of it (which corresponds to globally ordered structure) was produced by a rule including a small random term.

7. DISCUSSIONS

Symmetry concepts have neurobiological relevance both at all levels of the structural hierarchy (from molecules via synapses, networks and modules to the two hemispheres) and from functional point of view (e.g. perception).

The very essence of the neural is the emergence of order by self-organization (Szentágothai & Érdi, in press). The interplay between genetically given deterministic algorthm and random (on the macroscopic i.e. supramolecular level) spontaneous activity in the excitatory and inhibitory neurone populations lead to spatiotemporal order.

The problem of 'order - disorder - organization' - penetrated from thermodynamics to theoretical biology. (For a newer selection of papers on thermodynamics and biological pattern formation see Lamprecht & Zotin 1988). Biological - i.e. nonequilibrium systems in thermodynamic sense - structures are the results of 'symmetry breaking' instabilities. By symmetry breaking transitions the homogeneous, unstructured matter is organized into coherent, patterned structures. It was already Aharon Katzir Katchalsky, who asked in his seminal posthumus work (Katchalsky et al 1974), how the dynamic patterns of brain cell assemblies are formed by self-organizing, symmetry breaking mechansims. Further roles of the concepts of symmetry, asymmetry and dissymetry in the understanding and explanation of brain structures and functions might be discovered and/or analyzed in the near future.

ACKNOWLEDGEMENTS

I should like to thank to Prof. János Szentágothai his continuous moral support. Discussions with Prof. David Piggins on the problem of 'nonlinear seeing' during my short visit of the University of Guelph is highly acknowledged. Permissons from authors and copyright holders for republishing figures are highly appreciated.

REFERENCES

Anderson, P.A., Olavarria, J. & Van Sluyters, R.C,1988, J. Neurosci., 8:2183.
Barlow, H.B., 1969, Ann. N.Y. Acad. Sci. 156:872.
Barlow, H.B. & Foldiak, P., (in press),In:The Computing Neuron, Durbin, R. et al (eds), Addison-Wesley,N.Y.
Barlow, H.B. & Reeves., B.C., 1979, Vision Res.,19:783.
Barna, G. & Érdi, P, 1986,In:Cybernetics and Systems'86 Trappl, R. (ed.), D. .Reidel, Dordrecht, pp.343.
Black, M.M. & Baas, P.W., 1989, Trends in Neurosci. 12:211.
Blasdel, G.,G., & Salama, G., 1986, Nature, 321:579.
Braitenberg, V., 1986, in:Cerebral Cortex, Vol.3., Peters, A. & Jones, E.G.(eds.), Plenum, pp. 379.
Braitenberg, V. & Braitenberg, C., 1979, Biol. Cybern. 33:179.
Campbell, F.W.& Robson, J.G., 1968, J. Physiol. 197:551
Changeux, J.-P., 1985, Neuronal Man, Oxford Univ. Press New York - Oxford.
Changeux, J.-P. & Danchin, A., 1976, Nature, 264:705.
Eades, P., 1985, In:Computational Morphology, Toussaint, C.T. (ed.), Elsevier, Amsterdam, pp. 41.
Érdi, P. & Barna, G., 1984, Biol. Cybern. 51:93.
Érdi, P., & Barna, G., 1987, In:Mathematical Topics in Population Biology, Morphogenesis and Neurosciences, Teramoto, E. & Yamaguti, M, (eds.), Springer-Verlag, Berlin, pp.301.
Farkas, T.F. & Érdi, P., 1985, Leonardo, 18:179.
Ferster, D. & LeVay, S., 1978, J. Comp. Neurol. 182:923
Gierer, A. & Meinhardt, H., 1972, Kybernetik, 12:30.
Goldman, P & Nauta, W., 1977, Brain Res. 122:393.

Goetz, K.G., 1987, Biol. Cybern. 56:107.
Goetz, K.G., 1988, Biol. Cybern. 58:213.
Hargittai, I. & Hargittai, M., 1987, Symmetry through the Eyes of a Chemist, VHC Publ., New York.
Harrison, L.G., 1987, J. theor. Biol. 125:369.
Hebb, D.O., 1949, The Organization of Behavior, Wiley, New York.
Hoffman, W.C., 1966, J. Math. Psychol. 3:65.
Horsthemke, W. & Lefever, R., 1984 Noise-induced Transition, Springer-Verlag, Berlin.
Horton, J.C. & Hubel, D., 1981, Nature, 292:762.
Hubel, D. & Wiesel, T.N., 1959, J. Physiol. 148:574.
Hubel, D. & Wiesel, T.N., 1962, J. Physiol. 160:106.
Hubel, D. & Wiesel, T.N., 1968, J. Physiol. 195:215.
Hubel, D. & Wiesel, T.N., 1972, J.Comp. Neurol. 146:421
Hubel, D. & Wiesel, T.N., 1974, J.Comp. Neurol. 158:267
Hubel, D. Wiesel, T.N. & Stryker, M.P., 1978, J. Comp. Neurol. 177:361.
Katchalsky, A., Rowland, V. & Blumenthal, R., 1974, Dynamic Patterns of Brain Cell Assemblies, Neurosci. Res. Prog. Bull., Vol. 12.1.
Killackey, H.P., 1985.
Lamprecht, I. & Zotin, A.I. (eds), 1988, Thermodynamics and Pattern Formation in Biology, Walter de Gruyter New York.
LeVay, S., Hubel, D. & Wiesel, T.N., 1975, J. Comp. Neurol., 159:559.
Levy, J., 1988, In:Beauty and the Brain, Rentschler, I, Herzberger, B.& Epstein, D.eds, Birkhaeuser Verlag, Basel, pp. 219.
Livingstone, M.S., 1988, Sci. Am. pp. 68.
Livingstone , M.S. & Hubel, D. 1981, Nature, 291:554.
Locher, P.& Nodiner, C., 1989, Computers Math. Appls, 4-6: 475.
Maffei, L. 1977, In:Handbook of Sensory Perception, Held, R. et al, eds, Vol.VIII, Springer, pp.39.
Mallot, H.A., 1985, Biol. Cybern., 52:45.
Martin, K.A.C., 1988, Trends in Neurosci., 11:380.
May, R., 1972, Am. Nature 107:621
Meakin, P., 1986, J. theor. Biol. 118:101.
Mountcastle, V.B., 1957, J. Neurophysiol. 20:408.
Mower, G.D., Caplan, C.J., Christen, W.G. & Duffy, F.H., 1985, J. Comp. Neurol. 235:448.
Nobile, A. & Riccardi, L.M., 1984, Biol. Cybern. 49:177
Palm, G., 1982, In: Cybernetics and System Research, North - Holland, Amsterdam.
Peliti, L & Valpiani, A. (eds), 1988, Measures of Complexity, Springer, Berlin.
Pert, C.B. & Snyder, S., 1973, Proc. Natl. Acad. Sci. USA, 70:2243.
Ramon - Moliner, E. & Nauta, W.J.H., 1966, J. Comp. Neurol. 126:311.
Rakic, P., 1988, Science, 241:170.
Rentschler, I., Herzberger, B. & Epstein, D., 1988, Beauty and the Brain, Birkhaeuser-Verlag, Basel.
Rosen, R., 1981, Prog. Theoret. Biol. 6:161.
Schatz, C.J., Lindstrom, S. & Wiesel, T.N., 1977, Brain Res., 131:103.
Schierwagen, A., (in press),In:Organizational Constraints on the Dynamics of Evolution., (Vida,G Maynard Smith, J, eds, Manchester Univ Press.

Schwartz, E.L., 1977, Biol. Cyber. 28:1.
Schwartz, E.L., 1980, Vision Res. 20:645.
Seelen, W. von, Mallot, H.A. & Giannakopoulos,F, 1987, Biol. Cybern. 56:37.
Sepherd, G.M., 1983, Neurobiology, Oxford Univ. Press, New York - Oxford.
Slack, J.M., 1987, Trends in Biochem., 12:200.
Singer, W., 1987, In: The Neural and Molecular Bases of Learning, Changeux, J.-P. & Konishi, M.(eds), John Wiley & Sons Ltd., pp. 301.
Sperry, R., 1974, In:The Neurosciences:Third study program, Schmitt, F.F. & Worden, F.G. , eds.
Springer, S.P. & Deutsch, G., 1981, Left Brain, Right Brain, W.H. Freeman and Co, San Francisco
Stanley, H.E. & Ostrowsky, N. (eds.), 1986, On Growth and Form, Martinus Nijhoff Publ, Boston, Dordrecht.
Stryker, M.P. & Harris, W.A., 1986, J. Neurosci. 6:2117
Swindale, N.V., 1982, Proc R Soc Lond B., 215:211.
Szentágothai, J., 1978, Proc R Soc Lond B., 201:219.
Szentágothai, J., 1983, Rev. Physiol. Biochem. Pharmacol., 98:11.
Szentágothai, J., 1987, Adv. Physiol. Res., McLennan, M. et al, eds, Plenum:New York, pp. 111.
Szentágothai, J. & Arbib, M.A., 1974., Conceptual Models of Neural Organization, Neurosci. Res. Prog. Bull. 12.3.
Szentágothai, J. & Èrdi, P. (in press), J. Social Biol. Struct. 12:
Thom, R., 1975, Structural Stability and Morphogenesis, Benjamin Inc.: London.
Tootell, R.B., Silverman, M.S. & DeValois, R.L., 1981, Science, 214:813.
Tsonis, P.A., 1987, Trends in Biochem., 12:249.
Turing, A., 1952, Phil. Trans. Roy. Soc. 237:32.
Waessle, H., Peiche, L. & Boycott, B.B. 1981, Nature, 292:342.
Wiesel, T.N., 1982., Nature, 299:583.
Wolpert, L., 1969, J. theor. Biol. 25:1.
Wong - Riley, M.T.,1989, Trends in Neurosci., 12:94.
Zeki, S & Shipp, S., 1988, Nature 335:311.

THE EMERGENCE OF LIVING FUNCTIONS FROM MOLECULES

SIDNEY W. FOX

INSTITUTE FOR MOLECULAR AND CELLULAR EVOLUTION

UNIVERSITY OF MIAMI
MIAMI, FL U.S.A. 33177

ABSTRACT

A first experimental retracement of the transition from appropriate prebiotic matter (thermal protein) to polybiofunctional protocells is discussed. (Many details are necessarily presented in the references). The experiments suggest a molecular matrix that fits well with Schrodinger's "statistico-determinism." The primary source of information is inferred to be the molecular selection involved in an ancient system displayed by laboratory protocells. The phenomena are judged as more primitive than modern, but as a group they are quite comprehensive. The total model provides a basis for defining some of the problems of living systems relative to symmetry and symmetry-breaking.

ASYMMETRIC ORGANIC MOLECULES AND LIVING SYSTEMS

The symmetries within single playing cards and the symmetries within arrangements of them have played a dominant role in shaping thinking about the origin and transfer of biological information. This approach has been especially well illustrated by George Gamow in his 1955 Scientific American article on Information Transfer in the Living Cell (October 1955, p. 70-79). The assumed symmetries within and between amino acids and within and between monoribonucleotides are still widely held even though experiments in the late 1950s began to show that some of the assumptions were untenable. The preceptions of random arrangements of amino acids, compatible with symmetry, have begun to change (Fox, 1988b).

What has been missing from the Gamovian view was that amino acids are distinguishable by their stereochemical identities. As in Gamow's treatment (1955), the usual interpretation does not explain where the hereditarily crucial DNA of modern systems got its information, nor that the amino acids are themselves informational. By ignoring the roles of stereomolecular identity of the amino acids one ignores a fundamental chemical basis for all information in the living cell. Since 1955, bioevolutionary consequences deriving from this information base have been retraced in experiments; the holistic new overview is retraced in this paper.

In discussing these relationships from the perspective of symmetries, we are not referring to the enantiomeric symmetry of D and L amino acids, but to the whole-body stereoelectronic asymmetric identity of each of the sibling members of the amino acid family. While symmetry is in part

Symmetries in Science IV
Edited by B. Gruber and J. H. Yopp
Plenum Press, New York, 1990

suggested by the presence of an amino and a carboxyl group in each type of amino acid, the individual types are powerfully distinguished by their sidechains. As long as amino acids were themselves likened unto playing cards they were theoretically expected to combine in direct reactions by chance, i.e. randomly. By this reasoning, something like a genetic coding mechanism was needed at the outset to overcome the randomness and to provide specific arrangements of amino acids.

The experiments indicate, however, that the whole-body asymmetry of amino acids was the molecular basis for such arrangements of amino acid sequence, that these arrangements in the first proteins were biological information, and that this information evolved to the functional specificities in the modern living cell.

MOLECULAR EVOLUTION

The vast majority of theoretical retracements of the emergence of life from its precursors have been theoretical of an _a priori_ sort. This is to say that, in such cases, the student has assembled on paper a series of two or more chemical events that he believed could yield living systems. Experiments sometimes come later. The best known example of this _a priori_ approach is found in the works of Oparin (1957,1971).

After the subject area became primarily experimental in a cognitive way (Herrera, 1942; Fox and Dose, 1977), almost all experiments concerned the synthesis of biomonomers and of protobiopolymers from such biomonomers. Although the famous Urey-Miller experiment produced mainly amino acids in our context the synthesis called much attention to the larger problem of the origin of life (Miller, 1953).

Definitive experiments on thermal proteins and the even yet incomplete studies of polynucleotides (Fox and Dose, 1972; Miller and Orgel, 1974)

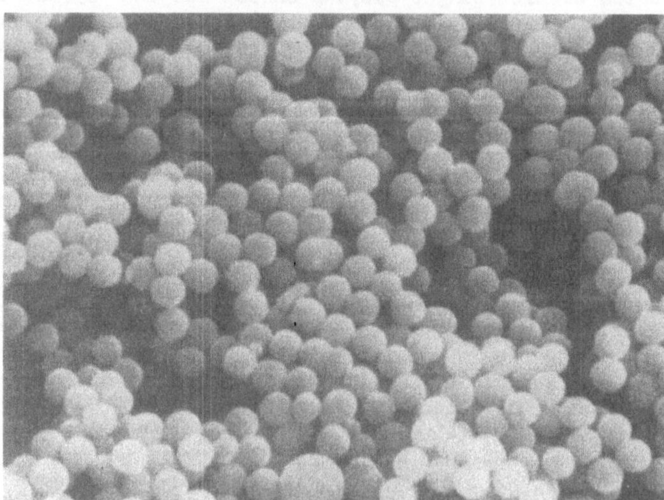

Figure 1. Scanning electron micrograph of proteinoid microspheres, models for protocells. These are usually mostly uniform, typically in one size 1-7 μm in diameter.

shifted attention to protobiopolymers. The need for the further and integrating stage of cell construction to fill out the picture has to a degree been recognized by experiments (Oparin, 1957; Fox et al. 1958; Yanagawa et al, 1980; Kenyon, 1984). However, few programs have begun at the molecular beginnings and then continued into cell construction (Young,1984). Herrera's (1942) was probably the first to do so.

Herrera produced amino acids, a pigment, a polymer, and cell-like structures--as primitive models and in a primitive scientific way. Herrera could not have been expected to have performed amino acid analysis by modern methods since they (Spackman et al, 1958) had not been developed. Even so, he identified glycine and cysteine; analysis by modern methods (Perezgasga, 1989) confirm those results and report also alanine and methionine. Herrera could not have been expected to have done any significant amount of amino acid sequence determination because that also did not get properly started (Fox, 1945; Rosmus and Deyl, 1972) until after Herrera's death in 1942. The inevitable controversies first concerned chemical aspects, such as whether first informational molecules were proteins or ribonucleic acids (Lehninger, 1975). Either of these two kinds of beginning, whichever is correct, needs to be justified by its leading to a biofunctional supramacromolecular system (Fox, 1988a). Any route that fails to do so is not pointed toward the origin of life (Fox, 1988a). Since the appearance of such a system in retracement experiments has been incisively claimed (Fox, 1980) the focus of attention has shifted to the biological end of the bridge that spans from inanimate matter to living systems. It is since Herrera's time that the selfsequencing of amino acids has been demonstrated (Fox, 1988), the selforganization to protocells has been explained (Fox, 1985), and the numerous biofunctional properties of proteinoid microspheres have been catalogued (Fox and Dose, 1977; Fox, 1988).

A generally acceptable definition of what is biological would be helpful in judging progress toward the goal of biological functionality but such definitions have typically, and understandably, been avoided by biologists. Biologists recognize that the living cell is an association of structures and functions. Experts are generally willing to characterize functions at the same time that they are unwilling to define life, and they have listed characteristics in a number of publications. In order to compare the functions of the proteinoid microsphere (Fox, 1985) with the characteristics of life as set forth in biological sources, we here use a recent textbook of biology (Wessells and Hopson, 1988).

Wessells and Hopson list as criteria of life: complexity, the taking in and use of energy, growth and development, reproduction, variation based on heredity. adaptation, and responsiveness.

Before analyzing this comparison in detail, a note of caution is in order. The focus of attention is in this paper on a protobiological organism rather than on a modern biological entity. To use the latter as a rigorous standard for a protoorganism is in one way like quietly endorsing instant creation. To compare the modern organism with the primordial organism, of which we do not have a sample in hand, however, requires new approaches such as stepwise synthesis, forward extrapolation, reliance on selforganization, etc. (Fox, 1988a).

CHARACTERISTICS OF CELLS AND OF SIMULATED PROTOCELLS

Complexity

The proteinoid microsphere is very complex in a molecular sense, although it arises from quite simple precursors. This fact is explained by the observation that simplicity modulates to complexity very rapidly (Fox, 1985). The rapid change to complexity is to be understood on the basis that the numbers of interactions between macromolecules of a few types proceed exponentially.

Inasmuch as the complex associations found are contained within single aspheroidal cells, we confront a perspective in which macromolecular complexity of a limited but significant kind is contained within cells that appear to represent a simple type.

Energy Absorption and Use

Proteinoids and proteinoid microspheres have been shown in two ways to use energy for metabolic purposes. In one, they convert each of various acids to their decarboxylation products much more rapidly in the presence of white light than in its absence (Wood and Hardebeck, 1972). The acids that have been studied (as the sodium salts) include pyruvate, glyoxalate, and glucuronate.

The other kind of reaction to have been studied was the phosphorylation of adenosine diphosphate, ADP->ATP (Fox et al, 1978). In each of these two studies energy was required for the conversions studied.

The photocatalyst for introduction of the needed energy has been found to be flavin and pteridine (Heinz et al, 1979), produced in the thermolysis of the amino acid mixtures used in generating the polypeptides. These pigments are evidently bound covalently into the polypeptide chains. They represent how a primordial type of photosynthesis could have originated prebiotically before a chlorophyll-based photosynthesis arose to be selected.

Growth and Development

Proteinoid microspheres grow to a predetermined size, dependent upon the identity of the polymer (Fox, 1988a). The growth is accretive, or heterotrophic. Again, this occurs without any DNA or RNA or similar materials in the history of the thermal protein that assembles itself. This kind of growth appears to differ from that of modern cells only by the fact that the modern type derives its structural protein from internal synthesis instead of in a preformed state from the environment (Fox et al, 1967).

The proteinoid microspheres demonstrate an innate tendency to develop. One of the manifestations of such tendency is found especially in microspheres assembled from hydrophobe-rich thermal proteins (Fox et al, 1987). Polymers of this class also stimulate the outgrowth of fibers of rat forebrain neurons in culture (Hefti et al, 1989); this finding further integrates prebiotic activity with modern evolution.

Association of two or more microspheres (Hsu et al, 1971) represents development of another kind.

Reproduction

Reproduction through a budding cycle resembles growth in accreting systems. When the individual microsphere attains its maximum size, which is programmed during accretive growth by the specific polymer, further deposition of thermal protein tends to appear on the microsphere as buds (Fox et al, 1967; Lehninger, 1975). Simulated binary fission, sporulation, and parturition have also been observed (Fox, 1973b).

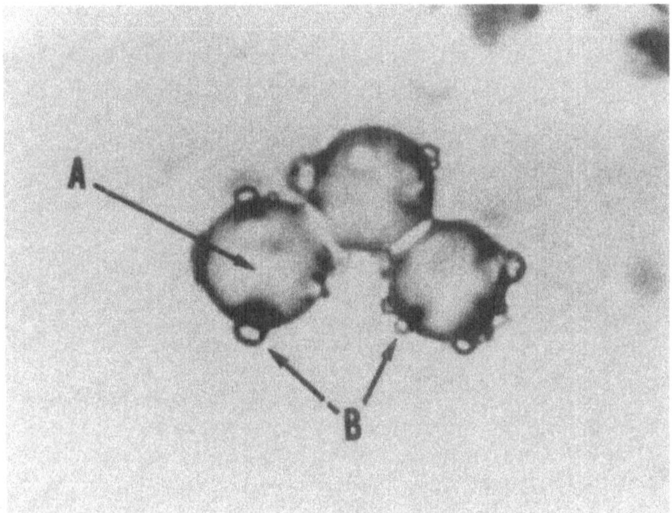

Figure 2a

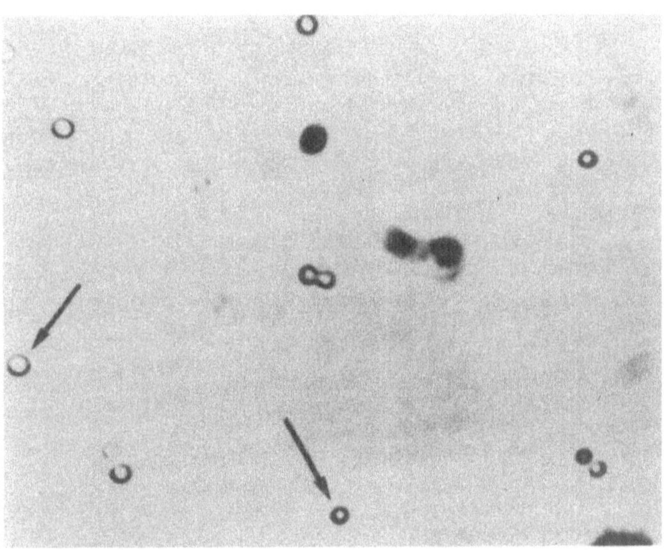

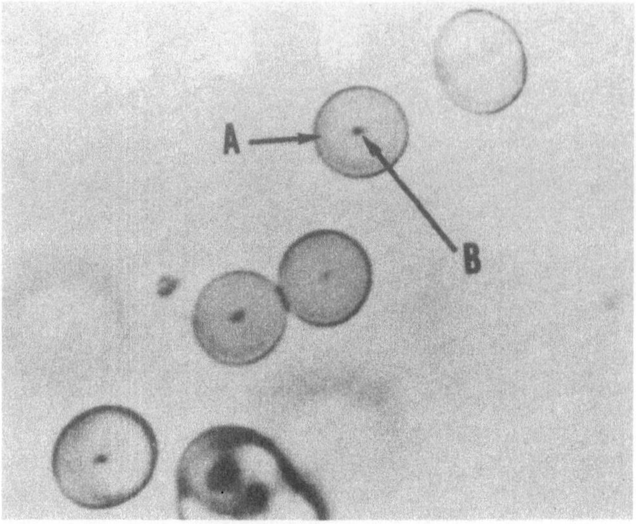

Figure 2c

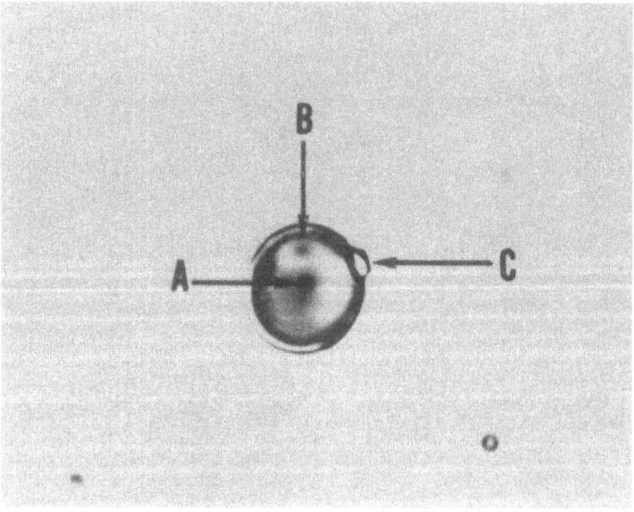

Figure 2d

Figure 2. Accretive reproduction of proteinoid microspheres. a. Buds on parent microspheres, b. liberated buds, c. microspheres grown from a central, darkly stained bud, d. mature second-generation microsphere with bud (Fox, 1973; Lehninger, 1975).

Variation Based on Heredity

The variation appears to be intrinsic and to be sustained within determined limits. This is true, at least, at the level of the initial proteins. The products from the selfsequencing of amino acids are close to homogeneous, but they are measurably heterogeneous. They comprise in any one synthesis a narrow range of types. Accordingly, the opportunity for internal variation existed at the outset. In the second generation, however, the range of types was not in the usual sense inherited from the parent; rather the range of types was what it was because of its assembly from the precursors that were much alike in both generations.

Modern heredity, in which DNA/RNA play key roles, was thus preceded in evolution by what was essentially a kind of endogenously controlled process of heredity, involving the selfsequencing of amino acids.

Adaptation

That living things are adapted to their environment is best understood in the large context that the first living things began in what was, strictly speaking, a pre-environment. The first living things came out of such a realm and can be said to have been continuous from the realm that preceded its first environment. Evolution has sometimes been characterized as the progressive effort of the organism to free itself from its environment, which it has never succeeded in doing.

Responsiveness

The proteinoid microspheres, like the cells of today, respond to heat, light, chemical substances, electricity, other microspheres (other organisms), and other influences that affect true cells (Fox, 1988a). Of special interest is the electrical behavior elicited by the action of light on proteinoid microspheres. Electrical recordings thus obtained are remindful of neurons and other excitable cells.

Wessells and Hopson summarize the relationship of the behavior of organisms to their history when they state,

> "Every living thing on Earth today is a descendant of an organism that lived before it. Each is a member of an unbroken lineage stretching backward in time to the era, billions of years ago, when life processes first became associated with organized sets of matter. Thus a knowledge of evolutionary history is important to our understanding of many characteristics of present-day organisms."

The accumulated knowledge acquired in retracing the emergence of the first organism now strongly supports this long view and the insight with which it was eloquently expressed. What the model emphasizes in addition is that life processes are not merely associated with, nor continuous with, other life processes; they are intrinsic manifestations of the appropriate matter.

In summary, we have a partially chemical, protobiological, and basically evolutionary answer to Schrödinger's 1946 question of What is Life? and the beginning of an answer to Delbrück's (1978) question of Mind from Matter?? Schrödinger preferred an answer of 'order-from-order' instead of 'order-from-disorder'. He traced his answer from Planck's discussions. It was not until some years after Delbruck's question that a detailed answer including the beginning of excitability, such as attaches to mind, could be articulated (Fox, 1980, 1984).

For protolife we perceive already an <u>association</u> of materials and processes. The concept of (unsymmetrical) order stems also from chemical evidence (Fox, 1988) and interdigitates with the newer inferences from the Big Bang (Fox, 1980), as in Barrow and Silk's "The Left Hand of Creation " (1983); see also Seiden (1986). The chemical view is a materialistic one by definition, but it has philosophical interfaces with physics and biology.

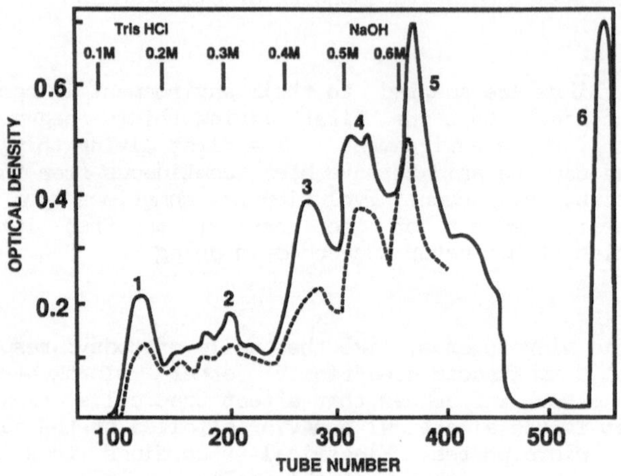

Figure 3a

MOLECULAR SELECTION

Without exception, the experiments in which mixtures of various amino acids have been brought together and the sequences that have been studied reveal that the amino acids sequence, or order, themselves. This was first seen in the difference of amino acid composition of the reaction mixture and of the polymer formed, as by heating (Fox and Harada, 1958). Some chemists would have expected such disparity between mixture and polymer, but others would not (cf. Fox and Windsor, 1984). The fact that the polymers have a small number of individual components separable by composition would not have signified that separation by sequence would reflect separation by composition. It however was found to do so when the experimental results were examined by sequence content (Nakashima et al,1977). By all the criteria used in all the various laboratories to put the question to the test, nonrandom peptide formation was found to be the result of thermal polycondensation of amino acids (Fox, 1980). Comparisons with modern proteins are feasible (Mikelsaar, 1976).

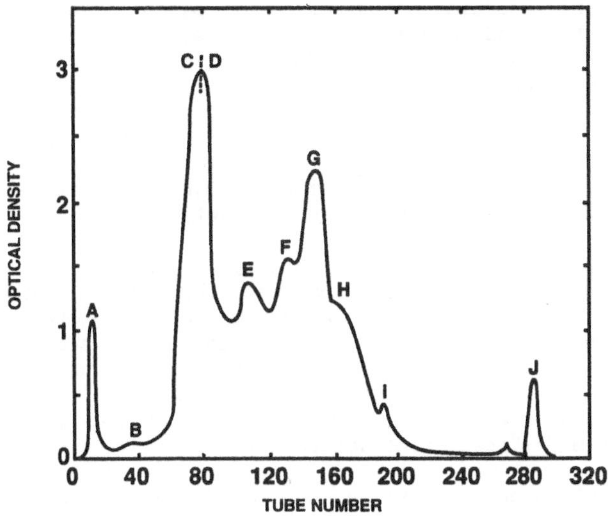

Figure 3b

Figure 3. a. Fractionation of amidated 1:1:1-proteinoid on DEAE-cellulose column. A few major fractions are observed. Repetitions yield similar patterns. The dotted line represents one of these. b. Limited heterogeneity in thermal proteinoid preparation and in turtle serum proteins, fractionated similarly (Fox and Nakashima, 1984).

The first studies of heterogeneity compared the total compositions of the polymers with the compositions of reaction mixtures and then with N-terminal or C-terminal analyses (Fox and Harada, 1960). These overwhelmingly indicated nonrandomness. More penetrating was the separation of components and carrying out such analyses on the components (Melius and Sheng, 1975). Subsequently, amino acid sequences in individual components were determined by the then widely used overlap technique (Fox, 1945).

It has been subsequently recognized that the production of amino acids by reactions of appropriate precursors also gives nonrandom arrays. With all of these results, we infer that production, polycondensation, and individual reactions of amino acids are individual for the same stereoelectronic reasons that amino acids undergo nonrandom polycondensation. In other words, all organic molecules contain information. How does this relate to symmetry?

THE PRESENT ERA OF BROKEN SYMMETRY

One of the consequences of an experimental laboratory model for the emergence of life is that it provides a vehicle for testing the experimental results against prevailing theory and vica versa. A first concept to have been tested in this way was that of randomness. In testing the experimental results against the a priori theory one first finds that there is no single theory. The relevant question of whether the matrix was random (Eigen, 1971) or nonrandom has in fact been a matter of dissension for decades. Although the problem is complicated by alternate terminologies, the question seems often to place on one side Einstein, Schrödinger, Bohm, and Pauling, for example, and on the other Bohr, Heisenberg, Dirac, Dyson et al. The question also raises a doubt of whether the right questions are being asked. In addition to divergent points of view, the seeking of a higher, or more unified, perspective is complicated by words that have different connotations for different students. Such words are uncertainty, indeterminism, determinism, random, and the like. Do the results , for example, comport better with the determinism of Einstein, or with the uncertainty principle of Heisenberg?

Is the basic problem one of causality vs. indeterminacy? Does the problem exist at a subatomic level and then recede at a higher level? The emphasis on Order out of Chaos (Prigogine and Stengers, 1984) is not supported by the protobiological experiments, which themselves began to move only when nonrandomness was recognized as pervasive reality (Fox, 1988). The results agreed better with Einstein's position in his three decades-long argument (Clark, 1971). They agreed yet better with Schrödinger's (1946) "statistico-determinism".

The cluster of concepts that are of immediate concern here come under the heading of symmetry. For a biochemist, assurance that symmetry is relevant to protobiology is found in the prologue of "The Left Hand of Creation" by two biophysicists, Barrow and Silk (1983). Their statement is,

> "Around us now, we see only left-handed neutrinos; their right-handed partners did not survive the early stages of the universe to emerge into the present era of broken symmetry. It is not only neutrinos that fail to be ambidextrous. Even biological development on earth is a breaking of perfect symmetry. All known DNA molecules, the building blocks of life, are left-handed spirals. We find only left-handed amino acids in living things , never their mirror images. The tale of broken symmetries extends from the beginning of time to the here and now".

This statement needs some editing. Some right-handed amino acids are indeed found in living things, although they are unusual. DNA molecules are blue-print carriers of inheritance instead of being building blocks of life. But biological development reflects a "breaking of perfect symmetry" if, indeed, biological development is not already well past the initial stage of symmetry-breaking.

EMERGENCE OF MIND

Some years ago, Per-Olov Löwdin stated, " You are not going to have a theory of the origin of life until you have also a theory of the origin of mind". In light of what I have learned since then this remark seems to me to have been a most appropriate one for a quantum scientist to make , especially one who has devoted efforts to the mind-matter relationship (Löwdin, 1986).

The essence of the concept of mind from matter is a recognition that decision-making is a process that can be found in matter (Matsuno, 1984; Mishra, 1984); decision-making need not wait until the mammalian brain has evolved from matter. The concept of decision-making in matter is quite precisely what we mean by molecular selection, i.e. by the selfsequencing of amino acids (Fox, 1986). It contrasts with the widely and deeply held tenet of randomness (cf. Fox and Windsor, 1984; Eigen,1971; Fox,1988a,Chap.9; Prigogine and Stengers, 1984). In this new view, the roots of mind could not be rigorously regarded as an integral aspect of matter and life until there had been an experimental demonstration of molecular selection interpreted in the context of evolution, i.e. of molecular evolution (Fox,1953).

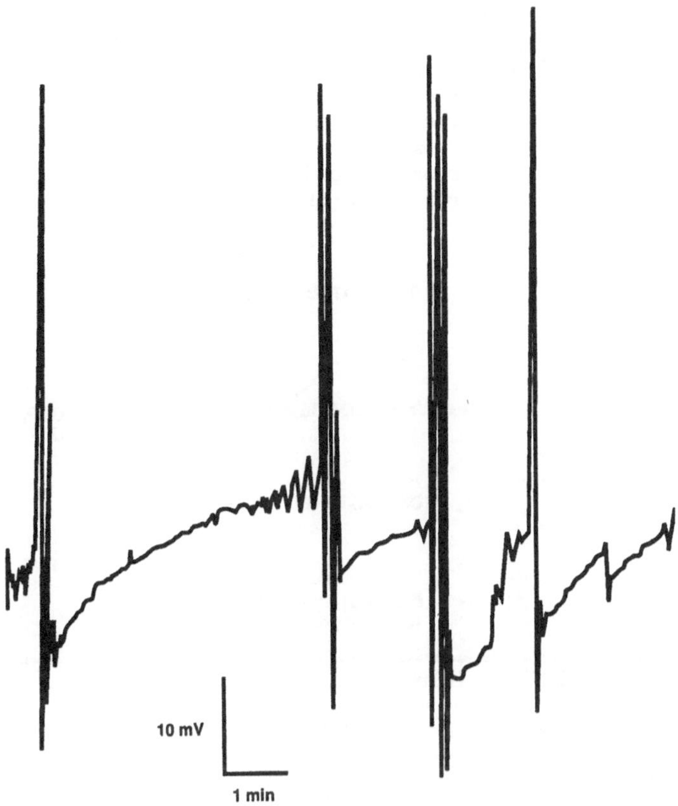

Figure 4. Action potentials of proteinoid microspheres resembling those in neurons (Przybylski and Fox, 1986).

EPILOG AND SUMMARY

The information detailed in a number of publications (Fox and Dose, 1977; Fox, 1988a) presents a connected and fairly comprehensive picture of how the selfsequencing of amino acids yielded informational (thermal) proteins in a geological environment. Ivanov and Förtsch (1986) report from computerized studies how the primordial selfsequencing of amino acids continued as a process through evolution into modern proteins. The evolution from molecular into cellular processes is retraced by the transition from inanimate matter to polybiofunctional cells by selforganization of thermal proteins, in turn informed by their precursor amino acids.

The thermal proteins organize to cellular structures and are found to have arrays of enzymatic activities (Rohlfing and Fox, 1969; Dose, 1984: Melius, 1982). The various microspheres studied have been found to have chemical and cytological behavior that qualifies as the roots of numerous other biofunctions as well. These simulate growth (programmed for cellular size), accretive reproduction, and membrane behavior (Fox et al, 1969) including associated electrical behavior (Fox, 1980 b; Przybylski and Fox, 1986; Fox, 1988a).

The experiments suggest also how a nucleic acid coding system could have come into existence from mononucleotides; the origin of the mononucleotides themselves is however yet to be adequately explained. Mononucleotides can, however, undergo polymerization catalyzed by thermal proteins that also catalyze peptide bond formation (Fox, 1981). Because of this dual activity in two kinds of synthesis by a single type of catalytic agent, the onset of a genetic coding mechanism from prior informational protein is visualized. While broken symmetry is perceived in biological development the breaking appears to have occurred early in evolutionary history.

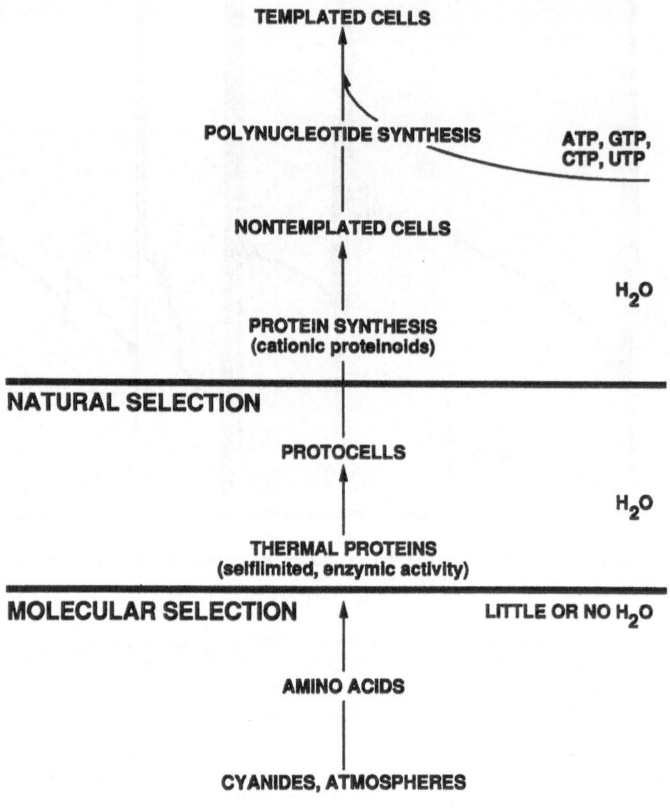

Figure 5. Flowsheet of molecular → cellular evolution (Fox, 1986).

REFERENCES

Barrow, J. S., and Silk, J., 1983, "The Left Hand of Creation", Basic Books, New York.

Delbrück, M., 1978, In "The Nature of Life", W. H. Heidcamp, ed., University Park Press, Baltimore, 140-166.

Dose, K., 1984, *Intl. J. Quantum Chem.,Quantum Biol. Symp.,* 11: 91.

Eigen, M.,1971, *Naturwissenschaften*, 58: 465.

Fox, S.W.,1945, *Advances Protein Chem.*, 2: 155.

Fox, S.W., 1973, *Pure Appl. Chem.*, 34:641.

Fox, S,W., 1980a, *Naturwissenschaften*, 67: 576.

Fox, S.W., 1980b, *Comp. Biochem. Physiol.*,67B: 423.

Fox, S.W., 1981, In "Science and Scientists", M. Kageyama, K.Nakamura, T. Oshima, and T. Uchida, eds. Japan Sc. Soc. Press, Tokyo, 39-45.

Fox, S.W., 1984, In "Beyond Neo-Darwinism, An Introduction to the New Evolutionary Paradigm", M.-W. Ho, and P.T. Saunders, eds., Academic Press, London, 15-60.

Fox, S.W., 1985, In "Structure and Motion: Membranes, Nucleic acids & Proteins", E. Clementi, G. Corongiu, M.H. Sarma & R. H. Sarma, eds., Adenine Press, Guilderland NY, 101-114.

Fox, S.W., 1986, *Quart. Rev. Biol.*, 61: 397.

Fox, S.W., 1988a, "The Emergence of Life, Darwinian Evolution from the Inside", Basic Books, New York.

Fox, S.W., 1988b, *FASEB J.* 2:2524.

Fox, S.W., and Dose, K., 1972, "Molecular Evolution and the Origin of Life", Freeman, San Francisco, revised edition, 1977, Marcel Dekker, New York.

Fox, S.W., and Harada, K., 1958, *Science*, 128: 1214.

Fox, S,W., and Harada, K., 1960, *J.Amer. Chem. Soc.*, 82: 3745.

Fox, S.W., and Middlebrook, M., 1954, *Federation Proc.*, 13: 211.

Fox, S.W. and Nakashima T., 1984, In "Individuality and Determinism", S.W. Fox, ed., Plenum Press, New York, 185-201.

Fox, S.W., and Windsor, C.R., 1984, *Intl. J. Quantum Chem.,Quantum Biol. Symp.* 11 :103.

Fox, S.W., Harada, K., and Kendrick, J.,1959, *Science*, 129: 1221.

Fox, S.W., McCauley, R.J., and Wood, A., 1967, *Comp. Biochem. Physiol.*, 20:773.

Fox, S.W., McCauley, R.J., Montgomery, P. O'B.,Fukushima, T., Harada, K., and Windsor, C.R., 1969, In "Physical Principles of Biological Membranes", F. Snell, J. Wolken, G.J.Iverson, and J. Lam, eds., Gordon and Breach, New York, 417-430.

Fox, S.W., Adachi,T., Stillwell, W., Ishima, Y., and Baumann, G., 1978, In " Light Transducing Membranes, Structure, Function, and Evolution", Academic Press, New York, 61-75.

Fox, S.W., Hefti, F., Hartikka, J., Junard, E., Przybylski, A. T., and Vaughan, G., 1987, *Intl. J. Quantum Chem., Quantum Biol. Symp.*, 14:347.

Hartmann, J., Brand, M.C., and Dose, K.,1981, *BioSystems*, 13: 141.

Heinz,B., Ried,W.,and Dose, K., *Angew.Chem.,Intl. Ed.Engl.*, 18: 478.

Herrera, A.L. 1942, *Science*, 96: 14.

Hsu, L.,L., Brooke, S., and Fox, S. W., 1971, *BioSystems* (Curr. Mod. Biol.) 4:12

Ivanov, O.C., and Förtsch, B., 1986, *Origins Life*, 17: 35.

Kenyon, D.H., 1984, In "Molecular Evolution and Protobiology", K. Matsuno, K. Dose, K. Harada, and D.L. Rohlfing, eds., Plenum Press, New York, 163-188.

Kerr, R.A., 1980, *Science*, 210: 42.

Lehninger, A.L., 1975, "Biochemistry", Worth and Co., New York.

Löwdin, P.-O., 1986, In "Selforganization", S.W. Fox, ed., Adenine Press, Guilderland NY, 23-33.

Matsuno, K.,1984, In "Molecular Evolution and Protobiology", K. Matsuno, K. Dose, K. Harada, and D. L. Rohlfing, eds., Plenum Press, New York, 433-464.

Melius, P., 1982, *BioSystems*, 15:275.

Miller, S.L., 1953, Science, 117:528.
Miller, S.L., and Orgel, L., 1974, "The Origins of Life on the Earth", Prentice Hall, Englewood Cliffs NJ.
Mishra, R. K.,1984, Intl. J. Quantum. Chem.,Quantum Biol. Symp.,11:45.
Nakashima. T., Jungck, J.R., Fox, S.W., Lederer, E., and Das, B.C., 1977, Intl. J. Quantum Chem., Quantum Biol. Symp., 4:65.
Oparin, A.I., 1957, "The Origin of Life on the Earth", 3rd ed., transl. by Ann Synge, Academic Press, New York.
Oparin, A.I.,1971, Sub-Cell. Biochem., 1:75.
Perezgasga, L., 1989, B.Sc. thesis, Univ. of Mexico, communicated by A. Negron and F.G. Mosqueira.
Prigogine, I., and Stengers, I., 1984, "Order out of Chaos", Bantam Books, New York.
Przybylski, A.T., and Fox, S.W., 1986, In "Modern Bioelectrochemistry", F.Gutmann, and H. Keyzer, eds., Plenum Press, New York, 377-396.
Rohlfing, D.L.,1976, Science, 193:68.
Rohlfing, D.L., 1984, In "Molecular Evolution and Protobiology", K. Matsuno, K. Dose, K. Harada, and D.L. Rohlfing, eds., Plenum Press, New York, 29-43.
Rohlfing, D.L., and Fox, S.W,, 1969, Advances Catal., 20: 373.
Rosmus, J., and Deyl, Z., 1972, Chromatogr. Rev.,70 :221.
Seiden, P.E., 1986, In "Selforganization", S.W. Fox, ed., Adenine Press, Guilderland NY, 1-21.
Spackman, D., Stein. W.H., and Moore, S., 1958, Anal. Chem., 30:1190.
Wessells , N.K., and Hopson, J.L., 1988, "Biology. Random House, New York, 3-7.
Wood, A., and Hardebeck, H.G., 1972, In "Molecular Evolution, Prebiological and Biological", D.L. Rohlfing, and A.I. Oparin, eds., Plenum Press, New York, 233-245.
Yanagawa, H., Kobayashi, Y., and Egami, F., 1980, J. Biochem., 87: 855.
Young, R.S., 1984, In "Molecular Evolution and Protobiology", K. Matsuno, K. Dose, K. Harada, and D. L. Rohlfing, eds., Plenum Press, New York, 45-48.

BIOMOLECULAR HANDEDNESS AND THE WEAK INTERACTION

A.J.MacDermott

Physical Chemistry Laboratory, University of Oxford

South Parks Road, Oxford OX1 3QZ

G.E.Tranter

Wellcome Research Laboratories, Wellcome Foundation

Beckenham, Kent BR3 3BS, England

Most biomolecules exist in two enantiomeric or mirror-image forms, but only one of these forms is used naturally in terrestrial biochemistry. This broken symmetry is now believed to be a feature of fundamental physics - a result of symmetry-breaking by the weak interaction. Animals are made of proteins built from L-amino acids, coded for by DNA built from D-sugars, to the virtual exclusion of the "unnatural" D-amino acids and L-sugars. We aim to show here that this is a result of "electroweak bioenantioselection": the L-amino acids and D-sugars are slightly more stable than their enantiomers due to the weak interaction. While aiming primarily at chemists, we include simple introductions to many of the necessary concepts in the hope of making this interdisciplinary review more accessible to biologists while retaining enough rigour for physicists.

The weak interaction is one of the four forces of nature. After a brief introduction in Section A, we discuss these four forces in section B, giving a simplified overview of concepts necessary for their description, such as metrics, Feynman diagrams, currents (derived for the Dirac equation by analogy with the more familiar Schrödinger equation) and propagators. In section C we concentrate on the weak interaction, which, unlike the other forces, can distinguish left and right. In section D we use the concepts of sections B and C to derive the weak interaction Hamiltonian

Symmetries in Science IV
Edited by B. Gruber and J. H. Yopp
Plenum Press, New York, 1990

which produces a small energy difference between left and right-handed molecules. In section E we see how this small energy difference can be amplified, and in section F we present the results of our calculations of this energy difference for some important biomolecules.

A. INTRODUCTION

Homochirality - the fact that all biomolecules are of one hand - is one of the hallmarks of life, being essential for an efficient biochemistry just like the universal adoption of right-handed screws in engineering. A homochiral pre-biotic chemistry is probably also a pre-requisite for the emergence of life because polymerization to form stereoregular biopolymers - such as poly-L-peptides - is found to proceed efficiently only in optically pure monomer solutions (Idelson and Blout 1958, Lundberg and Doty 1957, Blair et al 1981) because addition of the "wrong" enantiomer to the growing chain tends to terminate the polymerization. Moreover, molecules of the "wrong" hand often have destructive effects in living systems and must be eliminated by special enzymes such as D-amino acid oxidases. The classic example is of course thalidomide, in which one enantiomer was a useful drug and it was the mirror image that had the disastrous teratogenic effects.

The initial selection of one hand in ancestral biomolecules fixes the handedness of the rest of biochemistry through diastereomeric interactions (Fischer 1894). By way of illustration, a right hand in a right glove and a left hand in a left glove are <u>enantiomeric</u> or mirror-image systems, whereas a right hand in a right glove and a right hand in a left glove are <u>diastereomeric</u> systems: a right hand can feel the difference between right and left gloves. There is evidence for such a diastereomeric connection between the L-amino acids and the D-sugars (Wolfrom et al 1949, Melcher 1974), which would seem to preclude a D-amino acid/D-sugar or L-amino acid/L-sugar biochemistry. Mirror-image D-amino acid/L-sugar life should, however, be just as viable as natural L-amino acid/D-sugar life, so the question arises as to whether selection of the latter was a "frozen accident" (Miller & Orgel 1974, Cairns-Smith 1986) or, as Pasteur believed (Pasteur 1874, 1884a,b), the result of some universal chiral influence.

The universe is indeed pervaded by dissymmetry at many levels: Pasteur, who first established the connection between molecular and crystal dissymmetry in his famous resolution of tartaric acid crystals (Pasteur 1848), also noticed that the solar system is non-superimposable on its mirror image (Pasteur 1922), and it is now known that there is a predominance among spiral galaxies of rotation to the left with respect to the direction of recession (Borchkade & Kogoshvili 1976, Yamagata et al 1981, MacGillivray & Dodd 1985).

Pasteur thought the universal chiral influence might be a magnetic field, and later suggestions have included such supposedly chiral combinations as the Earth's magnetic and gravitational fields, or the Coriolis force and gravity. However, these influences can probably now be discounted because they are "falsely chiral" according to Barron's new definition of chirality (Barron 1986a,b, 1987a). A truly chiral system is one that exists in two distinct enantiomeric states that are interconverted by parity (space inversion, $x \rightarrow -x$, $y \rightarrow -y$, $z \rightarrow -z$) but not by time reversal ($t \rightarrow -t$). The hallmark of "true" chirality is natural optical rotation, which changes sign under parity but not under time reversal, that is, it is parity-odd, time-even. Absolute asymmetric synthesis can therefore be induced only by something with this same symmetry. A magnetic field will not do as it is parity-even (being an axial vector), time-odd (imagine it generated by a circular motion of electrons) and therefore falsely chiral. Collinear magnetic and electric fields also form a falsely chiral combination as the magnetic field is parity-even, time-odd, and the electric field parity-odd, time-even, giving overall a parity-odd, time-odd influence. Experiments claiming asymmetric synthesis in this way (Gerike 1975) - or analogously by stirring (parity-even, time-odd) in the Earth's gravitational field (parity-odd, time-even) (Dougherty 1980, Edwards et al 1980, Dougherty 1981) - can therefore probably be discounted. However it has recently been suggested (Barron 1986a, 1987a,b) that false chirality may be sufficient for asymmetric synthesis under conditions of kinetic rather than thermodynamic control, because the changing entropy then confers a preferred time direction which destroys any time-reversal symmetry (Birss 1966), but this view can be disputed (Mead & Moscowitz 1980).

The circularly polarized photon is a truly chiral influence with well-established enantioselective properties (Mason 1982, Mason 1983, Isumi & Tai 1977). Unfortunately, however, although sunlight is about 0.1% circularly polarized, it is equally and oppositely so at dawn and dusk (Angel et al 1972, Wolstencroft 1984), giving overall an even-handed influence - unless of course life started in a pool on an east facing mountain slope, but this is no more than a "frozen accident" explanation, as life could equally have started on a west-facing slope, with the opposite effect.

It would thus appear that most classical "chiral influences" are either falsely chiral or even-handed on a time or space average. But Pasteur was correct in thinking that there is a universal chiral influence, because it is now known that elementary particles themselves have an intrinsic handedness, felt only by the weak interaction.

B. THE FOUR FORCES

The weak interaction is one of the four forces of nature: the electromagnetic force is responsible for most of everyday chemistry and physics, the weak and strong forces are responsible for processes in the atomic nucleus, and gravity is familiar from everyday life.

Force	Electromagnetic	Weak	Strong	Gravity
Carried by	photon	$W^{\pm}$, Z^0	mesons, gluons	gravitons
Strength	1	10^{-11}	100	10^{-38}
Range	∞	10^{-15}cm	10^{-13}cm	∞

The four forces are believed to be really one force: Salam and Weinberg successfully unified the electromagnetic and weak forces (Weinberg 1967, Salam 1968), and the strong force is now being brought in with "grand unification" theories, but quantum gravity is proving more elusive. The world is made of spin-1/2 fermions (the leptons and their corresponding neutrinos, e^-, ν_e, μ^-, ν_μ, τ^-, ν_τ, and the quarks, u,d,s,c,t,b). These fermions interact by the four forces, which are clearly one because all four operate by a similar mechanism: the fermions exchange bosons with each other and the four forces differ only in the type of boson involved. Only one boson, the photon, carries the electromagnetic interaction, but three bosons, the W^+, W^- and Z^0 carry the weak interaction, and many types of meson and gluon carry the strong force.

Interactions are represented by Feynman diagrams of the type

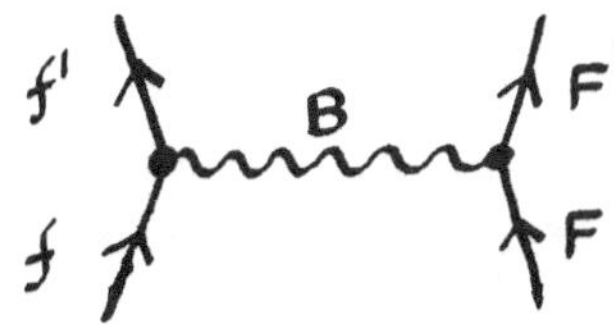

Here two fermions f and F are travelling up the page, and then a boson flashes between them across the page, causing the fermions to be deflected as they feel the force. The straight lines f and F represent the two fermions before they interact; the dots at the vertices represent interaction; the wiggly line represents the boson B that is exchanged; the straight lines f' and F' represent the fermions continuing on their way after interaction. In practice the fermions exchange not one but many bosons in the course of an encounter.

Examples are the electrostatic repulsion between two electrons, mediated by exchange of virtual photons, and the weak interaction between an electron and a neutron, mediated by virtual Z^0 bosons:

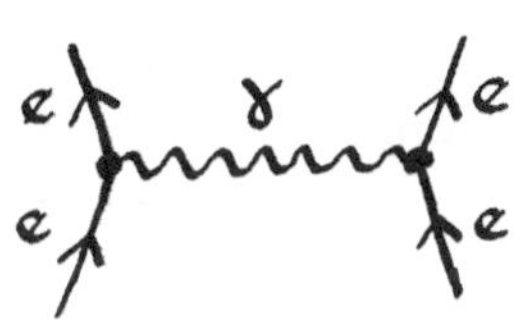

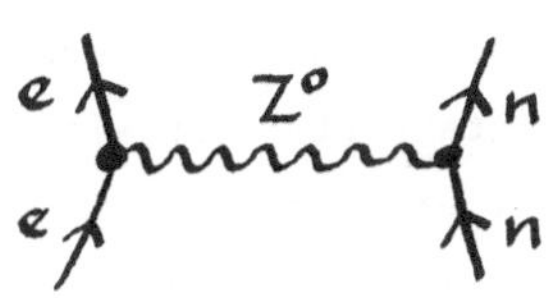

When two electrons repel each other by photon exchange one does <u>not</u> observe a beam of light flashing between them: this is because the exchanged photons are not real photons that can be detected, but "virtual" photons created from "borrowed" energy that must be "paid back" before they could be detected. This is because it takes a finite time to detect and measure a photon's energy, and from Heisenberg's energy-time uncertainty relation $\delta E\, \delta t \sim \hbar$ we see that almost any non-energy conserving process can happen provided the "borrowed" energy is "paid back" before it can be detected, i.e. within a very short time δt. Thus a low energy virtual radio photon could be created and survive for 10^{-6} s or more, while a high energy virtual γ-ray photon could only survive for about 10^{-20}s. Although virtual particles are not themselves detectable, their effects are certainly real and detectable, in much the same way as one could borrow a billion dollars and use it to change the world before paying it back a short time later.

In a Feynman diagram the final fermions f′ and F′ may differ from the initial fermions f and F in momentum (e.g. an electron may be slowed up by the attractive electrostatic pull of a proton) or even in flavour (e.g. in β-decay a down quark is converted by the weak force to an up quark). A momentum change can be appreciated by considering the analogy of two trains on parallel tracks, one moving faster than the other. If the windows are opened and the passengers throw oranges between the trains, momentum will be transferred and the faster train will be slowed down and the slower train speeded up. Furthermore, heavy cannonballs cannot be thrown as far as oranges, and this illustrates why elecromagnetism and gravity, mediated by massless photons and gravitons, have infinite range, while the weak and strong forces, mediated by very massive bosons, are very short range.

A flavour change can be understood if virtual bosons are allowed to momentarily become fermion/anti-fermion pairs: not only virtual energy, but also virtual matter can be momentarily created out of nothing provided it doesn't last very long and is in the form of particle/anti-particle pairs. Thus a virtual photon can momentarily become a virtual electron/positron pair: this is quite reasonable, as it is merely the reverse of the process by which electron and positron annihilate to give a γ-ray. Similarly the Z^0 boson of the weak interaction can also be considered as an electron/positron pair, or alternatively a neutrino/anti-neutrino pair, or quark/anti-quark pair. The W^+ and W^- bosons can exist as mixed flavour pairs: the W^- oscillates between down quark/anti-up quark ($d\bar{u}$) and electron/anti-neutrino ($e\bar{\nu}_e$) etc. while the W^+ oscillates between $u\bar{d}$, $\bar{e}\nu_e$ etc.. A down quark can therefore change into an up quark by emitting a W^-, or an up into a down by emitting a W^+:

Further details of the four forces can be obtained from Calder's (1976) pictorial and non-mathematical account.

To determine the energy associated with an interaction involving any of the four forces we employ a straightforward recipe to determine from the Feynman diagram the "Hamiltonian density" $\mathcal{H}(\vec{r})$ at a point $\vec{r}$, which is then integrated over all space to give the total energy. With each fermion we associate a "current", $J_\mu(ff')$ or $J^\mu(FF')$, which describes both initial (f) and final (f') fermion states. With each vertex we associate a coupling constant c representing the strength of the force involved. And with each virtual boson B we associate a "propagator" $\mathcal{P}_B$. The Hamiltonian density is simply the product of all these factors:

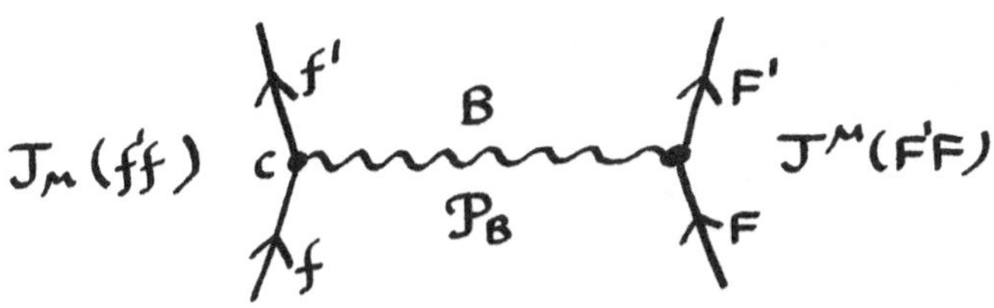

$$\mathcal{H}(\vec{r}) = J_\mu(f'f)\, c\, \mathcal{P}_B\, c\, J^\mu(F'F)$$

We shall first explain covariant and contravariant vectors - J_μ and J^μ - and for this we need to understand metrics. Then we shall derive expressions for currents and propagators, thus equipping ourselves to obtain the Hamiltonian density from any Feynman diagram.

<u>Metrics and covariant and contravariant vectors</u>

A metric is really nothing but Pythagoras' theorem. The distance s between two points (x',y',z') and (x,y,z) in ordinary 3-dimensional Euclidean space is given by Pythagoras theorem as

$$s^2 = (x'-x)^2 + (y'-y)^2 + (z'-z)^2$$

which can be written in differential form as

$$ds^2 = dx^2 + dy^2 + dz^2$$

or

$$ds^2 = g_{xx}dx^2 + g_{yy}dy^2 + g_{zz}dz^2$$

where the coefficients

$$g_{xx} = g_{yy} = g_{zz} = 1$$

are the components of the metric tensor

$$g_{\mu\nu} = \begin{pmatrix} 1 & 0 & 0 \\ 0 & 1 & 0 \\ 0 & 0 & 1 \end{pmatrix}$$

The zero off-diagonal elements indicate that terms in dxdy, etc. do not contribute to ds^2. Thus the components of the metric tensor indicate the weighting given to dx^2, dy^2, etc, in evaluating ds^2, the distance squared. In Euclidean space the weighting is the same for all coordinates since the x,y, and z directions are all equivalent. But there are other spaces in which different directions are not equivalent and therefore have different weightings. Pythagoras' theorem on the surface of a cylinder, for example, gives the metric

$$ds^2 = r^2 d\phi^2 + dz^2$$

or $g_{\mu\nu} = \begin{pmatrix} r^2 & 0 \\ 0 & 1 \end{pmatrix}$, i.e. $g_{\phi\phi} = r^2$

$$g_{zz} = 1$$

(z is the distance along the cylinder axis, ϕ is the angular coordinate). The weighting given to $d\phi$ in evaluating the distance depends on the radius: moving through a given angle $d\phi$ means a larger distance ds if the radius r is larger.

Returning to the three-dimensional Euclidean metric, we see that s^2, the squared length between two points, is independent of the coordinate axes used: the values of $(x'-x)^2$, etc. change, but their sum s^2 remains invariant under rotations, etc,. of the axes. According to relativity, not only lengths, but all physical laws must look the same in all coordinate frames. In particular, the speed of light c should appear the same to all observers, for if it did not, it would imply that there was some preferred frame of reference, which is clearly nonsense. This requirement provides a simple way to incorporate time into our metric - for relativity requires the fourth dimension, time, to be treated on an equal footing with the three space dimensions. If a pulse of light starts at the origin of coordinates, then the subsequent distance r of the pulse from the origin is given by

$$\frac{d|r|}{dt} = c, \qquad r = (x^2 + y^2 + z^2)^{1/2}$$

In another coordinate frame, in which the distance is

$$\bar{r} = (\bar{x}^2 + \bar{y}^2 + \bar{z}^2)^{1/2}$$

light still moves at speed c, so we have

$$\frac{d|\bar{r}|}{d\bar{t}} = c.$$

From these two equations we get

$$c^2 dt^2 - dr^2 = 0 = c^2 d\bar{t}^2 - d\bar{r}^2$$

We see that $c^2 dt^2 - dr^2$ is the same in all frames, so we can identify it with the invariant quantity ds^2, and write

$$ds^2 = c^2 dt^2 - dx^2 - dy^2 - dz^2$$

which is our new relativistic metric - the Lorentz metric. This throws new light on the meaning of c: it is just a conversion factor to convert time into the right units to be treated on an equal footing with distance. The Lorentz metric is often written

$$ds^2 = (dx^0)^2 - (dx^1)^2 - (dx^2)^2 - (dx^3)^2$$

or

$$g_{\mu\nu} = \begin{pmatrix} 1 & 0 & 0 & 0 \\ 0 & -1 & 0 & 0 \\ 0 & 0 & -1 & 0 \\ 0 & 0 & 0 & -1 \end{pmatrix}$$

in terms of the four coordinates (x^0, x^1, x^2, x^3), with $x^0 = ct$. The most striking feature of this metric is the negative sign of the spatial part: this is needed to forbid travel faster than light. For normal particles travelling slower than light we have $dr < cdt$, so $ds^2 > 0$. For luxons, particles travelling at the speed of light, we have $dr = cdt$ so $ds^2 = 0$. For a "tachyon", a particle travelling faster than light, we would have $dr > cdt$ or $ds^2 < 0$, which is obviously unacceptable since it implies imaginary distances.

The negative sign of the space components in the Lorentz metric has an important bearing on the difference between covariant and contravariant vectors. These are really the 4-dimensional analogue of row and column vectors. In the three dimensions of space we regard a scalar product as the product of a row vector and a column vector:

$$\vec{a}.\vec{b} = (a_x, a_y, a_z)\begin{pmatrix} b_x \\ b_y \\ b_z \end{pmatrix} = a_x b_x + a_y b_y + a_z b_z$$

If the fourth dimension of time is included then a scalar product is the product of a covariant vector and a contravariant vector:

$$A.B = \sum_{\mu=0}^{3} A_\mu B^\mu$$

Here Greek subscripts indicate the components of a covariant vector, A_μ, and Greek superscripts indicate the components of a contravariant vector, B^μ. To avoid cumbersome summation signs we adopt a notation in which summation over a repeated index is implied:

$$A.B = (A_0, A_1, A_2, A_3)\begin{pmatrix} B^0 \\ B^1 \\ B^2 \\ B^3 \end{pmatrix} = A_\mu B^\mu$$

Although covariant and contravariant vectors can for simplicity be viewed as row and column vectors, their rigourous definition is in terms of their properties under transformations of coordinates:

$$\bar{A}_\lambda = A_\alpha \left(\frac{\partial x^\alpha}{\partial \bar{x}^\lambda} \right) \quad \text{covariant}$$

$$\bar{B}^\lambda = B^\alpha \left(\frac{\partial \bar{x}^\lambda}{\partial x^\alpha} \right) \quad \text{contravariant}$$

Here summation over the repeated index α is implied, and $(\partial x^\alpha / \partial \bar{x}^\lambda)$ and $(\partial \bar{x}^\lambda / \partial x^\alpha)$ are matrices which transform "row" and "column" vectors respectively from the unbarred to the barred coordinate systems. The two matrices are clearly the inverse of each other, which is what one would expect: if a certain matrix is used to transform an ordinary column vector from one coordinate system to another, then the inverse of that matrix would be needed to effect the same transformation on a row vector.

Most "ordinary" vector quantities are contravariant, e.g. the 4-vector of infinitesimal increments (dx^0, dx^1, dx^2, dx^3):

$$d\bar{x}^{\mu} = \left(\frac{\partial \bar{x}^{\mu}}{\partial x^{\nu}}\right) dx^{\nu} \quad \text{contravariant.}$$

But vectors whose components are the derivative of some other quantity,

$$A_{\alpha} = \frac{\partial V}{\partial x^{\alpha}}$$

are covariant:

$$\frac{\partial V}{\partial \bar{x}^{\lambda}} = \left(\frac{\partial x^{\alpha}}{\partial \bar{x}^{\lambda}}\right)\left(\frac{\partial V}{\partial x^{\alpha}}\right) \quad \text{covariant}$$

The fact that the (covariant) derivative is in some sense the "inverse" of the (contravariant) infinitesimal increment again explains the use of respective transformation matrices that are the inverse of each other.

The covariant and contravariant components of a vector are related by

$$A_{\mu} = g_{\mu\nu} A^{\nu}$$

so the scalar product of two 4-vectors is

$$A.B = A_{\mu} B^{\mu} = g_{\mu\nu} A^{\nu} B^{\mu}$$

The metric tensor $g_{\mu\nu}$ is of course diagonal in the Lorentz metric, and we see that just as it weights the contribution of different dimensions to total distances, it also weights the contribution of different dimensions to scalar products.

In the Lorentz metric, 4-vectors have the form

$$A^\mu = (A^0, \vec{A}) \qquad \text{contravariant}$$

$$A_\mu = (A^0, -\vec{A}) \qquad \text{covariant}$$

where $\vec{A}$ is an ordinary vector in Euclidean 3-space, and A^0 is the "time-component". A typical contravariant 4-vector is the momentum 4-vector,

$$p^\mu = \left(\frac{E}{c}, \vec{p}\right)$$

(cf. the quantum mechanical operators for energy and momentum are respectively $i\hbar\, \partial/\partial t$ and $\hbar/i\, \vec{\nabla}$). Another is current, $j^\mu = (\rho, \vec{J})$, which we discuss in the next section. The derivative $\frac{\partial}{\partial x^\mu} = \partial^\mu$ is covariant (inverse of contravariant dx^μ) so it is written

$$\partial^\mu = (\partial^0, -\vec{\nabla}) = \left(\frac{1}{c}\frac{\partial}{\partial t}, -\vec{\nabla}\right)$$

Further information can be found in Atwater (1974), but the conclusions needed for the next section can be summarized as follows:

$$j^\mu = (\rho, \vec{J}), \quad j_\mu = (\rho, -\vec{J})$$

$$\partial^\mu = \left(\frac{1}{c}\frac{\partial}{\partial t}, -\vec{\nabla}\right), \quad \partial_\mu = \left(\frac{1}{c}\frac{\partial}{\partial t}, \vec{\nabla}\right)$$

$$j^\mu j_\mu = (\rho, \vec{J})\begin{pmatrix}\rho \\ -\vec{J}\end{pmatrix} = \rho^2 - \vec{J}\cdot\vec{J}$$

$$\partial^\mu \partial_\mu = \left(\frac{1}{c}\frac{\partial}{\partial t}, -\vec{\nabla}\right)\begin{pmatrix}\frac{1}{c}\frac{\partial}{\partial t} \\ \vec{\nabla}\end{pmatrix}$$

$$= \frac{1}{c^2}\left(\frac{\partial}{\partial t}\right)^2 - \nabla^2 = \Box$$

$$\partial^\mu \cdot j_\mu = \frac{1}{c}\frac{\partial}{\partial t}\rho + \vec{\nabla}\cdot\vec{J}$$

where $\Box$ is the D'Alembertian operator, a 4-dimensional generalization of the Laplacian operator ∇^2. We see that the negative sign of the space component of covariant vectors reflects the negative sign of the space components in the Lorentz metric. The signs can be confusing, as in $\partial^\mu \partial_\mu$ (product of covariant and contravariant vector) and $\partial^\mu j_\mu$ (product of two covariant vectors). But sign is very important in electroweak bioenantioselection because the sign of the weak Hamiltonian determines whether the L or the D enantiomers are more stable.

Fermion currents

"Current" is normally defined as the rate of flow of charge, i.e. charge x velocity. Now velocity is momentum/mass, so for electrons (charge -e), current is $e\vec{p}/m$. Current density is current weighted by the probability density $\psi\psi^*$, and therefore has the form $-(e/m)\psi^*\vec{p}\psi$. But as we require a real rather than imaginary quantity, we take this expression plus its complex conjugate to give the current density

$$\vec{J} = -\frac{1}{2}(e/m)\left(\psi^*\vec{p}\psi + \psi\vec{p}^*\psi^*\right)$$

But $\vec{p} = -i\hbar\vec{\nabla}$, so our final expression for the current density is

$$\vec{J} = i\hbar(e/2m)(\psi^*\vec{\nabla}\psi - \psi\vec{\nabla}\psi^*)$$

(Atkins 1983). This satisfies the "continuity equation",

$$\frac{\partial\rho}{\partial t} = -\nabla.\vec{J}$$

where $\rho = -e\psi\psi^*$ is the charge density. The continuity equation describes the flow of a charged fluid: clearly the rate of change of charge density is related to the current (rate of flow). If we are interested in the rate of flow not so much of charge as of probability of finding the particle, we divide by (-e) to give

$$\vec{J} = \frac{\hbar}{2im}(\psi^*\vec{\nabla}\psi - \psi\vec{\nabla}\psi^*)$$

for the "probability current". This still satisfies the continuity equation, but ρ is now the probability density $\psi^*\psi$, rather than the charge density.

The exact form of the probability current depends on which equation the wave functions ψ are solutions to. The current shown above is based on solutions to the Schrödinger equation - but here we shall require currents appropriate to the Dirac equation. A complete understanding of the Dirac equation is not necessary for our purposes - we simply need to be able to manipulate the currents derived from it. Full details of the Dirac equation are given in, for example, Aitchison and Hey (1988), Sakurai (1976) and we summarize here only the most important features.

The Schrödinger equation

$$i\hbar\frac{\partial\psi}{\partial t} = \left(-\frac{\hbar^2}{2m}\nabla^2\right)\psi$$

is derived from the non-relativistic energy-momentum relation

$$E = p^2/2m$$

by the prescription

$$E \to i\hbar\frac{\partial}{\partial t}, \quad \vec{p} \to \frac{\hbar}{i}\vec{\nabla}$$

The relativistic energy-momentum relation

$$E^2 = p^2c^2 + m^2c^4$$

leads to the Klein-Gordon equation,

$$-\hbar^2\frac{\partial^2\psi}{\partial t^2} = (-\hbar^2c^2\nabla^2 + m^2c^4)\psi$$

But if we use the approximate linearized form

$$E \approx pc + mc^2$$

we obtain the famous Dirac equation

$$i\hbar\frac{\partial\psi}{\partial t} = (-i\hbar c\vec{\alpha}.\vec{\nabla} + \beta mc^2)\psi$$

or

$$i\hbar\frac{\partial\psi}{\partial t} = -i\hbar c\left(\alpha_1\frac{\partial\psi}{\partial x^1} + \alpha_2\frac{\partial\psi}{\partial x^2} + \alpha_3\frac{\partial\psi}{\partial x^3}\right) + \beta mc^2\psi$$

But the correct relativistic energy-momentum relation must still be satisfied, i.e. the solutions to the Dirac equation must also satisfy the Klein-Gordon equation. This imposes conditions on the form of $\vec{\alpha}$ and β : it can readily be shown (by squaring both sides of the Dirac equation) that they must satisfy the anti-commutation relations

$$\{\alpha_i, \beta\} = 0$$

$$\{\alpha_i, \alpha_j\} = 2\delta_{ij}\mathbb{1}$$

also $\alpha_i^2 = \beta^2 = \mathbb{1}$

Two quantities A and B are said to commute if

$$AB - BA = [A, B] = 0$$

and anti-commute if

$$AB + BA = \{A, B\} = 0$$

Anticommutation is clearly not possible with ordinary numbers, so the α_i and β must be matrices. It can be shown that the Klein-Gordon equation is satisfied if

$$\alpha_i = \begin{pmatrix} 0 & \sigma_i \\ \sigma_i & 0 \end{pmatrix}, \quad i = 1,2,3, \quad \beta = \begin{pmatrix} \mathbb{1} & 0 \\ 0 & -\mathbb{1} \end{pmatrix}$$

where the σ_i are the Pauli matrices,

$$\sigma_x = \begin{pmatrix} 0 & 1 \\ 1 & 0 \end{pmatrix}, \quad \sigma_y = \begin{pmatrix} 0 & -i \\ i & 0 \end{pmatrix}, \quad \sigma_z = \begin{pmatrix} 1 & 0 \\ 0 & -1 \end{pmatrix}$$

These are used to describe the spin angular momentum $\vec{s}$ of a spin-1/2 fermion: we write $\vec{s} = \vec{\sigma}/2$ and the Pauli matrices are readily shown to satisfy the commutation relation

$$[\sigma_i, \sigma_j] = 2i\epsilon_{ijk}\sigma_k$$

that defines an angular momentum (cf. $[\hat{l}_x, \hat{l}_y] = i\hbar\hat{l}_z$, $[\hat{s}_x, \hat{s}_y] = i\hat{s}_z$, Atkins 1983) as well as the anti-commutation relation

$$\{\sigma_i, \sigma_j\} = 2\delta_{ij}\mathbb{1}$$

The Pauli matrices operate on "spinors" of the form $\chi = \begin{pmatrix} a \\ b \end{pmatrix}$

The eigenspinors of $\hat{s}_z$ are $\phi^1 = \begin{pmatrix} 1 \\ 0 \end{pmatrix}$ and $\phi^2 = \begin{pmatrix} 0 \\ 1 \end{pmatrix}$ corresponding to spin eigenvalues m_s=+1/2 and -1/2, i.e. "spin-up" and "spin-down":

$$\hat{s}_z \begin{pmatrix} 1 \\ 0 \end{pmatrix} = \frac{1}{2}\sigma_z \begin{pmatrix} 1 \\ 0 \end{pmatrix} = \frac{1}{2}\begin{pmatrix} 1 & 0 \\ 0 & -1 \end{pmatrix}\begin{pmatrix} 1 \\ 0 \end{pmatrix} = \frac{1}{2}\begin{pmatrix} 1 \\ 0 \end{pmatrix}$$

$$\hat{s}_z \begin{pmatrix} 0 \\ 1 \end{pmatrix} = \frac{1}{2}\sigma_z \begin{pmatrix} 0 \\ 1 \end{pmatrix} = \frac{1}{2}\begin{pmatrix} 1 & 0 \\ 0 & -1 \end{pmatrix}\begin{pmatrix} 0 \\ 1 \end{pmatrix} = \frac{1}{2}\begin{pmatrix} 0 \\ -1 \end{pmatrix} = -\frac{1}{2}\begin{pmatrix} 0 \\ 1 \end{pmatrix}$$

Note that ϕ^1 and ϕ^2 are not eigenspinors of σ_x or σ_y, i.e. only the z-component of the spin is determined, as one would expect. The Pauli matrices are Hermitian:

$$\sigma_q^\dagger = \tilde{\sigma}_q^* = \sigma_q$$

and the Dirac matrices $\vec{\alpha}$ and β are therefore also Hermitian. (A matrix M is Hermitian if it is equal to its Hermitian conjugate $M^\dagger$ (Atkins 1983). To take the "Hermitian conjugate" $M^\dagger$ of a matrix M, we take the transpose of the matrix - exchange rows and columns - to give $\tilde{M}$, and take the complex conjugate of the elements (replace i by -i) to give $\tilde{M}^* = M^\dagger$)

The appearance of the Dirac equation can be improved if we define the γ-matrices, which are components of a 4-vector $\gamma^\mu = (\gamma^0, \vec{\gamma})$, where

$$\gamma^0 = \beta \quad \text{("time-component")}$$

$$\gamma^k = \beta\alpha_k \quad \text{("space-components" of } \vec{\gamma}\text{)}$$

Notice that γ^0 is Hermitian, but the γ^k are anti-Hermitian:

$$(\gamma^0)^\dagger = \gamma^0, \quad (\gamma^k)^\dagger = -\gamma^k$$

This difference in sign is a very important one, and we have presented the details of these matrices in order to point it out, for getting this sign right has an important bearing on whether it is the L or the D enantiomers which are calculated to be more stable. Sakurai (1976) and Dirac (1958) define the γ^k so that they are Hermitian, $\gamma^k = -i\beta\alpha_k$, while Aitchison and Hey (1988) define them so that they are anti-Hermitian, $\gamma^k = \beta\alpha_k$ as above. The Hamiltonian (section C) for the weak interaction that differentiates between enantiomers has opposite sign when expressed in terms of γ-matrices defined by the two different conventions, and appalling confusion can therefore arise if it is not appreciated which convention is being used.

We will stick to anti-Hermitian γ^k here: it appears fitting that the space components show Hermitian behaviour of opposite sign to that of the time components, for this ties in with the Lorentz metric discussed earlier. The structure of this metric - opposite signs for space and time - is also reflected in the commutation relations

$$\{\gamma^\mu, \gamma^\nu\} = 2g^{\mu\nu}\mathbb{1}$$

The Dirac equation

$$i\hbar c\frac{\partial\psi}{\partial x^0} = (-i\hbar c\vec{\alpha}.\vec{\nabla} + \beta mc^2)\psi$$

(where we have put $x^0 = ct$) becomes, on multiplication through by β,

$$i\hbar c\beta\frac{\partial\psi}{\partial x^0} = (-i\hbar c\beta\vec{\alpha}.\vec{\nabla} + \mathbb{1}mc^2)\psi$$

which we recognize as

$$i\hbar c\gamma^\mu\psi_{,\mu} = mc^2\psi$$

where $\psi_{,\mu} = \partial\psi/\partial x^\mu$. This most elegant form of the Dirac equation treats all four dimensions on an equal footing.

The solutions to the Dirac equation for electrons and positrons are written

$$\psi(e^-) = N\,u(p,s)\,e^{-(i/\hbar)p.x}$$

$$\psi(e^+) = N\, v(p,s)\, e^{+(i/\hbar)p\cdot x}$$

where

$$u(p,s) = \begin{pmatrix} \phi^s \\ \dfrac{c\vec{\sigma}\cdot\vec{p}}{E+mc^2}\phi^s \end{pmatrix}, \quad v(p,s) = \begin{pmatrix} \dfrac{c\vec{\sigma}\cdot\vec{p}}{E+mc^2}\chi^s \\ \chi^s \end{pmatrix}$$

N is a normalization factor, and

$$p^\mu = \left(\frac{E}{c}, \vec{p}\right), \quad x^\mu = (ct, \vec{r}), \quad E = (p^2c^2 + m^2c^4)^{1/2}.$$

ϕ and χ are both 2-component spinors, with $\phi^1 = \begin{pmatrix}1\\0\end{pmatrix}$, $\phi^2 = \begin{pmatrix}0\\1\end{pmatrix}$

for spin up and down electrons respectively, and $\chi^1 = \begin{pmatrix}0\\1\end{pmatrix}$, $\chi^2 = \begin{pmatrix}1\\0\end{pmatrix}$

for spin up and down positrons respectively. Further details are to be found in Aitchison and Hey (1988), Sakurai (1976) and Dirac (1958).

To get the probability current for a given wave equation, the following procedure is used: starting from the equation and its complex conjugate, we multiply the equation by ψ^*, and its complex conjugate by ψ ; the two equations are then subtracted, and the expression tidied up and made to look like the continuity equation, so that the current can then be identified. Thus, for the Schrödinger equation we have

$$\psi^* i\hbar \frac{\partial \psi}{\partial t} = \psi^*\left(\frac{-\hbar^2}{2m}\nabla^2\right)\psi$$

$$-i\hbar\psi \frac{\partial \psi^*}{\partial t} = \psi\left(\frac{-\hbar^2}{2m}\nabla^2\right)\psi^*$$

Subtracting and dividing through by $i\hbar$ gives

$$\left(\psi^*\frac{\partial\psi}{\partial t} + \psi\frac{\partial\psi^*}{\partial t}\right) + \frac{\hbar}{2mi}\left(\psi^*\nabla^2\psi - \psi\nabla^2\psi^*\right) = 0$$

which is the same as

$$\frac{\partial \rho}{\partial t} + \nabla . \vec{J} = 0$$

if

$$\rho = \psi^* \psi, \quad \vec{J} = \frac{\hbar}{2mi}(\psi^* \nabla \psi - \psi \nabla \psi^*)$$

as obtained earlier.

For the Dirac equation we can obtain the current by an analogous process, but as ψ is a vector and not a scalar one must consider the Hermitian rather than complex conjugate, and reverse the directions of the operators:

$$i\hbar \psi^\dagger \frac{\partial \psi}{\partial t} = \psi^\dagger(-i\hbar c \vec{\alpha} . \vec{\nabla} + \beta mc^2)\psi$$

$$-i\hbar \frac{\partial \psi^\dagger}{\partial t} \psi = \psi^\dagger(i\hbar c \vec{\alpha} . \overleftarrow{\nabla} + \beta mc^2)\psi$$

Subtracting and dividing through by $i\hbar$ gives

$$\left(\psi^\dagger \frac{\partial \psi}{\partial t} + \frac{\partial \psi^\dagger}{\partial t}\psi\right) + c\left(\psi^\dagger \vec{\alpha} . \vec{\nabla} \psi + \psi^\dagger \vec{\alpha} . \overleftarrow{\nabla} \psi\right) = 0$$

or

$$\frac{\partial}{\partial t} \psi^\dagger \psi + c \nabla . (\psi^\dagger \vec{\alpha} \psi) = 0$$

which is the same as $\frac{\partial \rho}{\partial t} + \nabla . \vec{J} = 0$ if

$$\rho = \psi^\dagger \psi, \quad \vec{J} = c \psi^\dagger \vec{\alpha} \psi$$

This should be compared with the probability current for the Schrodinger equation, which had the form $\psi^* \frac{\vec{p}}{m} \psi$: $\vec{p}/m$ is a velocity, and in the Dirac current this is replaced by $c\vec{\alpha}$ which is also a velocity.

The Dirac probability current can be written as a 4-vector,

$$j^\mu = c \bar{\psi} \gamma^\mu \psi = (c\psi^\dagger \psi, c\psi^\dagger \vec{\alpha} \psi)$$

where $\bar{\psi} = \psi^\dagger \gamma^0, \quad \gamma^0 = \beta, \quad \gamma^k = \beta\alpha_k,$

if we take $x^0 = ct$ so that the continuity equation is

$$\partial_\mu j^\mu = 0 .$$

If, in a Feynman diagram, the fermions are not the same before and after interaction, the probability current becomes the fermion transition current

$$j^\mu(f'f) = c\,\bar{\psi}_{f'}\gamma^\mu \psi_f$$

or

$$j^\mu(f'f) = N_{f'}N_f\,\bar{u}_{f'}\gamma^\mu u_f\, e^{(i/\hbar)(p'-p)\cdot x}$$

Propagators

The propagator is the term associated with the boson exchanged during an interaction. To describe an interaction, a potential $\hat{V}$ is added to the Hamiltonian in the familiar Schrödinger equation (e.g. the electrostatic interaction between electron and proton in the case of the hydrogen atom). A similar procedure is used in the Dirac equation,

$$i\hbar c\frac{\partial\psi}{\partial x^0} = (-i\hbar c\vec{\alpha}\cdot\vec{\nabla} + \beta mc^2 + \hat{V})\psi$$

and the form of $\hat{V}$ can be deduced, in the case of electromagnetic interactions, by making the replacement

$$\partial^\mu \longrightarrow \partial^\mu - \frac{ie}{\hbar c}A^\mu$$

and identifying the extra terms with $\hat{V}$ to give

$$\hat{V} = (-e)(A^0\mathbb{1} - \vec{\alpha}\cdot\vec{A})$$

This perturbation may induce transitions, and from time-dependent perturbation theory (Atkins 1983), the probability amplitude for such transitions is

$$a_{f'f} = \frac{1}{i\hbar} \iint d^3x\, dt\, \psi_{f'}^{*}\, \hat{V}\, \psi_f$$

Using the above potential, we readily obtain

$$a_{f'f} = \frac{-e}{i\hbar} \int d^4x\, \bar{\psi}_{f'} \gamma_\mu A^\mu \psi_f$$

We then recognize $j_\mu(f'f) = (-e)c\,\bar{\psi}_{f'}\gamma_\mu \psi_f$

as the fermion transition current, giving

$$a_{f'f} = \frac{1}{i\hbar c} \int d^4x\, j_\mu(f'f) A^\mu$$

This expression shows that an interaction can be represented as

(fermion current) x (potential A^μ due to other particles)

The "other particles" are not only the other fermion, with which interaction is taking place, but also the boson exchanged during the interaction. We earlier stated that an interaction described by a Feynman diagram had the form

(fermion current) x (propagator) x (other fermion current).

Therefore the potential has the form

A^μ = (propagator) x (other fermion current),

so if we can find A^μ, we can then idenfify the propagator.

For electromagnetic interactions $A^\mu = (\Phi, \vec{A})$, where Φ and $\vec{A}$ are the scalar and vector potentials which generate the electric and magnetic fields through

$$\vec{E} = -\nabla\Phi - \partial\vec{A}/\partial t$$

$$\vec{B} = \nabla\wedge\vec{A}$$

The electric and magnetic fields are related to any charges ρ and currents $\vec{J}$ by the four Maxwell equations

$$\nabla.\vec{E} = \rho$$

$$\nabla\wedge\vec{E} = -\frac{1}{c}\frac{\partial\vec{B}}{\partial t}$$

$$\nabla.\vec{B} = 0$$

$$\nabla\wedge\vec{B} = \vec{J} + \frac{1}{c}\frac{\partial\vec{E}}{\partial t}$$

Using $A^{\mu} = (\Phi, \vec{A})$ and $j^{\mu} = (\rho, \vec{J})$, the Maxwell equations can be neatly expressed as

$$\Box A^{\mu} - \partial^{\mu}(\partial_{\nu}A^{\nu}) = j^{\mu}$$

where we recall that the D'Alembertian operator is given by

$$\Box = \partial_{\mu}\partial^{\mu} = \frac{1}{c^2}\frac{\partial^2}{\partial t^2} - \nabla^2$$

We shall first find the propagator for virtual photons. Here we may work in the Lorentz gauge, $\partial_{\nu}A^{\nu} = 0$, so that the Maxwell equations become

$$\Box A^{\mu} = j^{\mu}(F'F)$$

where $j^{\mu}(F'F)$ is the current for the other fermion F with which our fermion f is interacting. To solve for A^{μ}, we note that the current

$$j^{\mu}(F'F) = \bar{u}_{F'}\gamma^{\mu}u_F\, e^{(i/\hbar)(p'-p).x}$$

carries all its x-dependence in the exponential. Now since

$$\Box e^{iq.x} = -q^2 e^{iq.x}$$

we see that

$$A^\mu = -q^{-2} e^{iq.x} \quad \text{or} \quad -q^{-2} j^\mu(F'F)$$

is a solution (where q = p′ - p is the momentum transfer). We can therefore identify the propagator:

photon propagator: $-\frac{1}{q^2}$

To find the propagator for the massive bosons exchanged in the weak interaction, we take a generalized form of the Maxwell equations (Lorentz qauge is applicable only to massless bosons), making the substitution $\Box \longrightarrow \Box + M^2$, where M is the mass of the boson (cf. the Klein-Gordon equation):

$$(\Box + M^2) W^\mu - \partial^\mu(\partial_\nu W^\nu) = j^\mu$$

As before, we solve for W^μ, the potential for the weak interaction, and realize that this is the propagator times the current, so the propagator can be picked out as

$$\mathcal{P}_B(\text{massive}) = \frac{-g^{\mu\nu} + q^\mu q^\nu / M^2}{q^2 - M^2}$$

Now the momentum transfer is very small compared to the huge mass of the exchanged boson (for the W bosons, M ~ 90 GeV), so this readily reduces to:

massive boson propagator: $\frac{1}{M^2}$

In summary, therefore, the Hamiltonian density is

$$\mathcal{H}(F) = J_\mu(f'f)\, e \left(-\frac{1}{q^2}\right) e\, J^\mu(F'F)$$

for an electromagnetic interaction involving exchange of a massless photon, and

$$\mathcal{H}(F) = J_\mu(f'f)\, g \left(\frac{1}{M^2}\right) g\, J^\mu(F'F)$$

for a weak interaction involving exchange of a boson of mass M. The coupling constant for the electromagnetic interaction is the electronic charge e. The coupling constant g for weak interactions is actually not small, but similar in magnitude to e: it is the huge mass of the vector bosons, appearing squared in the denominator, which causes weak interactions to be weak.

C. THE WEAK INTERACTION

The weak interaction is the only one of the four forces which can tell the difference between left and right. Fermions exist in two states of opposite handedness or chirality, which are interconverted by parity (Aitchison & Hey 1988, Huang 1982). In a simple picture these can be viewed as states of opposite helicity, defined as $\lambda = \vec{s}.\vec{p}/|\vec{p}|$, corresponding to spin and momentum vectors parallel (right helicity, λ =+1) or anti-parallel (left helicity, λ =-1). One can define projection operators (Aitchison & Hey 1988),

$$P_R = \frac{1+\gamma_5}{2}, \quad P_L = \frac{1-\gamma_5}{2}$$

where $\gamma_5 = i\gamma^0\gamma^1\gamma^2\gamma^3 = \begin{pmatrix} 0 & 1 \\ 1 & 0 \end{pmatrix}$

so that

$$P_R = \frac{1}{2}\begin{pmatrix} \mathbb{1} & \mathbb{1} \\ \mathbb{1} & \mathbb{1} \end{pmatrix}, \quad P_L = \frac{1}{2}\begin{pmatrix} \mathbb{1} & -\mathbb{1} \\ -\mathbb{1} & \mathbb{1} \end{pmatrix}$$

When applied to a solution

$$u(p,s) = \begin{pmatrix} \phi^s \\ \dfrac{c\vec{\sigma}.\vec{p}}{E+mc^2}\,\phi^s \end{pmatrix}$$

of the Dirac equation, they project out its right and left handed parts:

$$P_R u = u_R = \begin{pmatrix} \phi^s + \dfrac{c\vec{\sigma}.\vec{p}}{E+mc^2}\,\phi^s \\[2ex] \phi^s + \dfrac{c\vec{\sigma}.\vec{p}}{E+mc^2}\,\phi^s \end{pmatrix}$$

$$P_L u = u_L = \begin{pmatrix} \phi^s - \dfrac{c\vec{\sigma}.\vec{p}}{E+mc^2}\,\phi^s \\[2ex] -\phi^s + \dfrac{c\vec{\sigma}.\vec{p}}{E+mc^2}\,\phi^s \end{pmatrix}$$

For massless particles the right and left-handed states u_R and u_L are eigenstates of the helicity operator $(1/2)\vec{\sigma}.\hat{p}$ (where $\hat{p}$ is a unit vector in the direction of the momentum $\vec{p}$). This can be shown readily by realising that for m = 0 we have

$$\frac{c\vec{\sigma}.\vec{p}}{E+mc^2} = \frac{c\vec{\sigma}.\vec{p}}{pc} = \vec{\sigma}.\hat{p}$$

and that since

$$\vec{\sigma}.\hat{p} = \begin{pmatrix} p_z & p_x - ip_y \\ p_x + ip_y & -p_z \end{pmatrix}$$

we have $(\vec{\sigma}.\vec{p})^2 = p^2 \mathbb{1}$ or $(\vec{\sigma}.\hat{p})^2 = \mathbb{1}$

This simple picture, in which right and left-handed particles have opposite helicity, breaks down for particles with non-zero mass, for u_R and u_L are then no longer eigenstates of the helicity operator, as we no longer have $\frac{c\vec{\sigma}.\vec{p}}{E+mc^2} = \vec{\sigma}.\hat{p}$. To see this physically one must appreciate that a massive particle can be overtaken by a moving observer, for its velocity must always be less than c: so a particle which appeared to have spin and momentum vectors parallel (right helicity) when the observer was behind it will appear to have spin and momentum vectors antiparallel (left helicity) when the observer has overtaken it, for it now appears to the observer to be going in the opposite direction. However, a massless particle, such as the neutrino, travels at the velocity of light and therefore cannot be overtaken and thus has the same helicity to all observers. Although u_R and u_L are not exact eigenstates of the helicity operator for massive particles, they <u>are</u> eigenstates of γ_5, the chirality operator, with eigenvalues +1 and -1 respectively. Chirality and helicity are exactly the same for massless particles, but not quite exactly the same for massive particles (Huang 1982). It is still a useful simplification, however, to think of right and left handed fermions as having spin and momentum respectively parallel and antiparallel, provided we realise that this picture is only exact in the limit of zero mass.

Since $P_R + P_L = 1$, it is clear that normal fermions can be viewed as a "racemic mixture", with equal numbers of the left and right-handed forms. This is analogous to a plane-polarized light beam, which can be viewed as a superposition of equal amounts of the right and left-circularly polarized forms (corresponding to spin-states $m_s = \pm 1$, with spin parallel or antiparallel to the direction of motion): there are always equal numbers of left and right-handed photons unless the beam is produced under the influence of chiral crystals, etc..

The two chirality states of fermions participate equally in the parity-conserving electromagnetic, strong and gravitational interactions, which therefore do not feel the handedness of fermions, seeing them as "racemic". However the two chirality states do not participate equally in the weak interaction, which therefore sees fermions not as "racemic" but as more one hand than the other. Parity is therefore violated in the weak interaction since it treats mirror images unequally: a left-handed electron participates in the weak interaction preferentially compared with a right-handed electron. But although P (parity) is violated, CP (parity plus charge conjugation) is conserved: left-handed fermions are preferred over right-handed fermions in the weak interaction, but the corresponding right-handed anti-fermions are preferred over left-hand anti-fermions to the same degree. As far as the weak interaction is concerned, therefore, particles and anti-particles have opposite handedness, and the dissymmetry that pervades the universe is the fact that it is made of matter, rather than anti-matter.

Barron's definition of true chirality has therefore been extended (Barron 1986b) so that true enantiomers are interconverted not by P, but by CP. Thus, the true enantiomer of a left-handed electron is not a right-handed electron, but a right-handed positron. There are numerous consequences of a true mirror being CP rather than just P. One is that all atoms and achiral molecules are predicted to be very slightly optically active owing to the handedness of their constituent elementary particles (Sandars 1980, Fortson & Wilets 1980). This very small "weak optical activity" has been detected for Bi atoms (Emmons et al 1983, 1984). Similarly, left and right-handed chiral molecules are not true enantiomers, but diastereoisomers: the true enantiomer of an L-amino acid is the D-amino acid made of anti-matter (Barron 1981). Left and right-handed molecules should therefore differ slightly in many properties, e.g. NMR chemical shifts (Barra et al 1987).

These differences between enantiomers are due to what might be termed the left-handedness of their constituent electrons. But it must be emphasized that left and right-handed electrons are normally present in equal numbers and the electron is only left-handed in the sense that its left chirality state participates preferentially in the weak interaction, which therefore sees electrons as mainly left-handed. However, where electrons are produced as a result of a weak process, there will be an actual excess of left-handed over right-handed electrons, in proportion to v/c, the velocity of the electrons relative to the velocity of light. Neutrinos are only ever produced as a result of a weak process, since they do not feel the electromagnetic and strong forces; and as they are massless they travel at the velocity of light so that $v/c = 1$ giving a 100% bias towards left-handedness. Neutrinos are therefore always left-handed: no right-handed neutrinos exist. Similarly, anti-neutrinos are always right-handed. Photons, however, can be both right and left-handed, because the photon (a boson) is its own anti-particle.

An example of a parity-violating weak process is radioactive β-decay, in which neutrons decay via the weak interaction into protons, anti-neutrinos and electrons:

$$n \rightarrow p + e^{-} + \bar{\nu}_e$$

As a neutron is two down quarks and an up quark (udd), whereas a proton is two ups and a down (uud), this is really the decay of a down quark (charge -1/3) into an up (charge +2/3) by emission of a W^- boson, which then decays into an electron and an anti-neutrino:

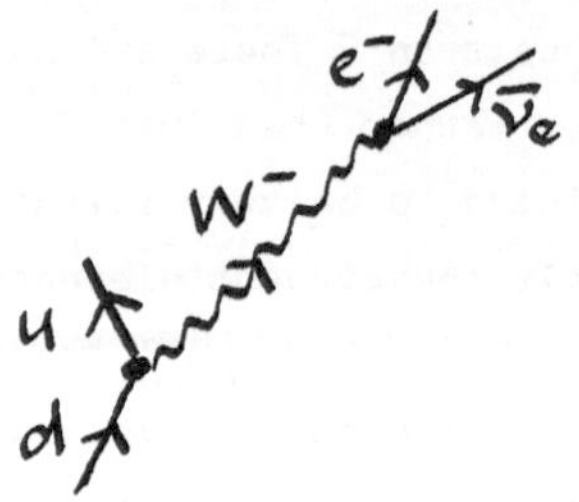

The electrons produced have predominantly left-handed spin polarization (Wu et al 1957). This excess of left-handed β-rays raises the possibility of enantioselective β-radiolysis as a symmetry-breaker in pre-biotic chemistry. In the laboratory differential absorption by chiral molecules of beams of left and right helically polarized electrons has been observed (Campbell & Farago 1985), but unfortunately there are problems in extrapolating this result to bioenantioselection in nature. β -decay electrons are produced at relativistic velocities, and in order to have an enantioselective rather than indiscriminately destructive effect they must be decelerated to lower energies. This is done very carefully in the laboratory, to preserve the helicity, but in nature deceleration involves scattering processes which are likely to spoil the helicity. In fact no reproducible bioenantioselective effects have been obtained experimentally (Bonner 1984, Van House et al 1984) from either β-radiolysis by the β-electrons themselves or β-photolysis by the circularly polarized bremsstrahlung radiation from their deceleration - but in theory the maximum enantiomeric excess obtainable could be as large as 10^{-12} (Hegstrom et al 1985, Hegstrom 1987), which, as will be shown later, is readily amplifiable to homochirality.

All the possible chiral influences discussed so far - β-emitters, circularly polarized light, magnetic fields, etc. - are local. A global and everpresent chiral influence is afforded by the fact that because left and right-handed molecules are really diastereoisomers, not enantiomers, they differ slightly in energy (Rein 1974, Letokhov 1975, Zel'dovich et al 1977). This parity-violating energy difference (PVED) between enantiomers arises from weak neutral current interactions, mediated by the Z^0 boson, between the (left-handed) electrons and the chiral nuclear framework. These interactions impart a parity-violating energy shift (PVES), E_{pv}, to the energy of a chiral molecule, and an equal and opposite shift, $-E_{pv}$, to that of its enantiomer, giving a parity-violating energy difference (PVED) of $\Delta E=2E_{pv}$.

Since the PVED reflects the "left-handedness" of the electron, by which we really mean the preference of the weak neutral current interaction for left-handed electrons, we need to know the extent of this preference. It is only neutrinos which are 100% left-handed in this respect; for other particles the extent to which the right and left-handed forms participate in the weak interaction has to be determined experimentally. Using the projection operators P_R and P_L we can express the fermion currents in the form (Aitchison & Hey 1988)

$$j_\mu(\nu) = c^\nu \bar{u}_\nu \gamma_\mu \left(\frac{1-\gamma_5}{2}\right) u_\nu$$

neutrino neutral current $\nu = \nu_e, \nu_\mu, \nu_\tau \ldots$

$$j_\mu(\ell) = \bar{u}_\ell \gamma_\mu \left\{ c_L^\ell \left(\frac{1-\gamma_5}{2}\right) + c_R^\ell \left(\frac{1+\gamma_5}{2}\right) \right\} u_\ell$$

charged lepton neutral current $l = e^-, \mu^-, \tau^- \ldots$

$$j_\mu(q) = \bar{u}_q \gamma_\mu \left\{ c_L^q \left(\frac{1-\gamma_5}{2}\right) + c_R^q \left(\frac{1+\gamma_5}{2}\right) \right\} u_q$$ quark neutral current $q = u, d, s, c, b, t \ldots$

where $c_L \neq c_R$ expresses the parity-violating nature of the weak interaction. The values of the coefficients are found to be

$c^\nu = 1/2$ for all ν

$c_L^l = -1/2 + a$, $c_R^l = a$ for all l

$c_L^q = 1/2 - 2a/3$, $c_R^q = -2a/3$ for u,c,t (all charge +2/3)

$c_L^q = -1/2 + a/3$, $c_R^q = a/3$ for d,s,b (all charge -1/3)

where $a = \sin^2 \Theta_W$ is a parameter in which Θ_W is known as the Weinberg angle. Experiment indicates that $\sin^2 \Theta_W = 0.23$, very close to the theoretical value of 0.25, corresponding to a Weinberg angle of 30°.

Vertices in weak neutral current (WNC) interactions are represented by the coupling constant g_N, related to the weak charged current (WCC) coupling constant by

$$g_N = \frac{g}{\cos \Theta_W}$$

The mass of the Z^0 boson is related to that of the $W^\pm$ bosons by

$$\cos \Theta_W = \frac{M_{W^\pm}}{M_{Z^0}}$$

The Hamiltonian densities for WNC and WCC interactions therefore have the form

$$j_\mu \, g_N \frac{1}{M_{Z^0}^2} g_N \, j^\mu \qquad \text{WNC}$$

$$j_\mu \, g \frac{1}{M_W^2} g \, j^\mu \qquad \text{WCC}$$

Now

$$g_N \frac{1}{M_{Z^0}^2} g_N = \frac{g^2}{\cos^2\theta_W} \frac{\cos^2\theta_W}{M_W^2} = \frac{g^2}{M_W^2}$$

so the Weinberg angle has cancelled out to give

$$\mathcal{H}(F) = \frac{g^2}{M_W^2} j_\mu j^\mu = \frac{8G_F}{\sqrt{2}} j_\mu j^\mu$$

where, by convention, $$\frac{G_F}{\sqrt{2}} = \frac{g^2}{8M_W^2}$$

where $G_F = 1.435 \times 10^{-62}$ Jm^3 = 2.222×10^{-14} a.u. is the "weak interaction coupling constant".

We are now equipped to calculate the Hamiltonian density for any weak interaction, and in the next section we use this to obtain an expression for the PVED.

D. DERIVATION OF AN EXPRESSION FOR THE PVED

In this section we first derive an expression for the Hamiltonian density from Feynman diagrams of the WNC interactions within a molecule. Next we reduce this expression to non-relativistic quantum mechanics, by explicitly multiplying out all the γ-matrices and spinors to give a parity-violating Hamiltonian operator. We then obtain the PVED by taking the appropriate matrix element of this Hamiltonian between molecular orbital wave functions. By using the LCAO approximation this reduces to a sum of atomic orbital matrix elements, and the final task is to show how this can be evaluated using the GAUSSIAN computer packages and our own PVED84 program. Further details of the derivation are in Tranter (1984), Mason & Tranter (1984,1985), and MacDermott & Tranter (1989).

Derivation of the Hamiltonian density

Since electrons, protons and neutrons all feel the WNC interaction, there are many such interactions to be considered. But since the weak interaction is a contact interaction (i.e. occurs only at very close range since the Z^0 is so heavy), interactions between nucleons in different nuclei can be ruled out. Furthermore, since there is no chirality inside a nucleus (the nuclei in left and right-handed molecules are obviously identical: the chirality of the molecule comes from the chiral mutual disposition of the nuclei), intranuclear interactions among the nucleons will not contribute to the PVED either. This leaves electron-electron, electron-proton, and electron-neutron interactions.

We therefore need the electron, proton and neutron weak neutral currents. For electrons, we have

$$j_\mu(e) = \bar{e}\gamma_\mu\left\{c_L^e\left(\frac{1-\gamma_5}{2}\right) + c_R^e\left(\frac{1+\gamma_5}{2}\right)\right\}e$$

where for simplicity we have written e, $\bar{e}$ instead of u_e ,$\bar{u}_e$.
Using $c_L^e = -1/2 + a$, $c_R^e = a$, and taking the theoretical value of the Weinberg angle to give $a = \sin^2\Theta_W = 1/4$, we obtain

$$j_\mu(e) = \bar{e}\gamma_\mu\left\{-\frac{1}{4}\left((1-4\sin^2\Theta_W)-\gamma_5\right)\right\}e$$

or

$$j_\mu(e) = \frac{1}{4}\,\bar{e}\gamma_\mu\gamma_5 e$$

For protons (uud) and neutrons (udd), we must examine the up and down quark currents separately, and by using the given values of c_L and c_R we obtain

$$j_\mu(u) = \frac{1}{4}\,\bar{u}\gamma_\mu\left\{1-\frac{8}{3}\sin^2\Theta_W-\gamma_5\right\}u$$

$$j_\mu(d) = \frac{1}{4}\,\bar{d}\gamma_\mu\left\{-1+\frac{4}{3}\sin^2\Theta_W+\gamma_5\right\}d$$

We then write $j_\mu(p) = 2j_\mu(u) + j_\mu(d)$

and $$j_\mu(n) = j_\mu(u) + 2j_\mu(d)$$

to give

$$j_\mu(p) = \frac{1}{4}\bar{p}\gamma_\mu\{(1-4\sin^2\theta_W)-\gamma_5\}p = -\frac{1}{4}\bar{p}\gamma_\mu\gamma_5 p$$

$$j_\mu(n) = \frac{1}{4}\bar{n}\gamma_\mu(\gamma_5-1)n$$

To contribute to the PVED, an interaction

$$\mathcal{H}(\vec{r}) = \frac{8G_F}{\sqrt{2}} j_\mu(f) j^\mu(F)$$

between fermions f and F must be parity-violating, i.e. it must change sign under the operation of parity, or inversion of coordinates. Under inversion of coordinates, momentum (a polar vector) changes sign but spin (an axial vector) does not, so a fermion wave function changes as follows:

$$u(p,s) = \begin{pmatrix} \phi^s \\ \frac{c\vec{\sigma}\cdot\vec{p}}{E+mc^2}\phi^s \end{pmatrix} \xrightarrow{\text{parity}} \begin{pmatrix} \phi^s \\ \frac{-c\vec{\sigma}\cdot\vec{p}}{E+mc^2}\phi^s \end{pmatrix}$$

This suggests that the parity operator is

$$\mathbb{P} = \begin{pmatrix} 1 & 0 \\ 0 & -1 \end{pmatrix} = \gamma_0$$

By evaluating $\gamma_0\gamma_\mu\gamma_0$ and $\gamma_0\gamma_\mu\gamma_5\gamma_0$ it can readily be shown that currents of the form $\bar{u}\gamma_\mu u$ are polar 4-vectors V_μ for which

$(V_0, \vec{V}) \longrightarrow (V_0, -\vec{V})$ under parity, while currents of the form $\bar{u}\gamma_\mu\gamma_5 u$ are axial 4-vectors A_μ for which $(A_0, \vec{A}) \longrightarrow (-A_0, \vec{A})$ under parity.

The PVED, like the optical rotation, is a pseudoscalar, i.e. it changes sign under parity, and must therefore be a product of a polar vector (parity-odd) and an axial vector (parity-even). If we take the theoretical value of the Weinberg angle, the electron and proton currents have only axial vector parts: the Hamiltonians for electron-electron or electron-proton interactions will therefore not be parity-violating. The only parity-violating contribution to the molecular WNC Hamiltonian therefore comes from the (axial vector) electron current and the polar vector part of the neutron current, giving

$$\mathcal{H}^{pv}(\vec{r}) = \frac{8G_F}{\sqrt{2}} j_\mu(e) j^\mu(n) = \frac{8G_F}{\sqrt{2}} \left(\frac{1}{4}\bar{e}\gamma_\mu\gamma_5 e\right)\left(-\frac{1}{4}\bar{n}\gamma_\mu n\right)$$

or

$$\mathcal{H}^{pv}(\vec{r}) = \frac{-G_F}{2\sqrt{2}} \bar{e}\gamma_\mu\gamma_5 e \bar{n}\gamma^\mu n$$

(Hegstrom et al 1980, Bouchiat & Bouchiat 1974a,b 1975) for the diagram

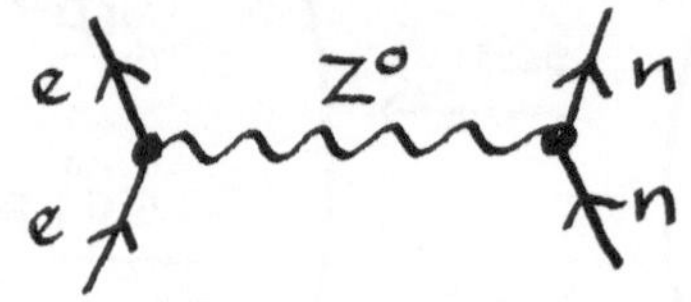

In order to get the correct sign for $\mathcal{H}^{pv}(\vec{r})$, we should recall at this point that we are using the convention

$$\gamma_5 = i\gamma^0\gamma^1\gamma^2\gamma^3, \quad \gamma^0 \text{ Hermitian}, \quad \gamma^{1,2,3} \text{ anti-Hermitian}$$

Some authors (Bouchiat & Bouchiat 1974a,b, 1975) use the convention

$$\gamma_5 = \gamma^0\gamma^1\gamma^2\gamma^3, \quad \gamma^{0,1,2,3} \text{ Hermitian}$$

and therefore obtain the opposite sign for $\mathcal{H}^{pv}(\vec{r})$.

Reduction to non-relativistic quantum mechanics

The quantum field theory expression

$$\mathcal{H}^{PV}(\vec{r}) = \frac{-G_F}{2\sqrt{2}}\, \bar{e}\gamma_\mu\gamma_5 e\, \bar{n}\gamma^\mu n$$

for the parity-violating Hamiltonian density at a point $\vec{r}$ becomes, on transformation to relativistic quantum mechanics, the parity-violating energy density

$$E^{PV}(\vec{r}) = \frac{-G_F}{2\sqrt{2}} \sum_i \sum_b \bar{e}\gamma_\mu(i)\gamma_5(i) e\, \bar{n}\gamma^\mu(b) n$$

where e,n are now relativistic multi-particle wave functions (at point $\vec{r}$) and the $\gamma_\mu(i)$ are one-particle operator operating only on particle i. The sum is over all electrons i and all neutrons b.

To reduce to non-relativistic quantum mechanics, we write

$$e = \begin{pmatrix} \phi_e \\ \chi_e \end{pmatrix}, \quad n = \begin{pmatrix} \phi_n \\ \chi_n \end{pmatrix}$$

where

$$\chi_e = \frac{c\vec{\sigma}\cdot\vec{p}}{E+m_e c^2}\phi_e, \quad \chi_n = \frac{c\vec{\sigma}\cdot\vec{p}}{E+m_n c^2}\phi_n$$

and use the γ -matrices

$$\gamma^\mu = (\gamma^0, \vec{\gamma}), \; \gamma_\mu = (\gamma_0, -\vec{\gamma}) = (\gamma^0, -\vec{\gamma})$$

explicitly, examining the "time" and "space" components separately. Thus, for example,

$$\bar{\psi}\vec{\gamma}\psi = \psi^\dagger\gamma^0\vec{\gamma}\psi$$

$$= (\phi^\dagger, \chi^\dagger)\begin{pmatrix}1 & 0\\ 0 & -1\end{pmatrix}\begin{pmatrix}0 & \vec{\sigma}\\ -\vec{\sigma} & 0\end{pmatrix}\begin{pmatrix}\phi\\ \chi\end{pmatrix}$$

$$= (\phi^\dagger\vec{\sigma}\chi + \chi^\dagger\vec{\sigma}\phi)$$

Similarly

$$\bar{\psi}\gamma^0\psi = \phi^\dagger\phi + \chi^\dagger\chi$$

$$\bar{\psi}\vec{\gamma}\psi = \phi^\dagger\vec{\sigma}\chi + \chi^\dagger\vec{\sigma}\phi$$

$$\bar{\psi}\gamma_0\gamma_5\psi = \phi^\dagger\chi + \chi^\dagger\phi$$

$$\bar{\psi}\vec{\gamma}\gamma_5\psi = \phi^\dagger\vec{\sigma}\phi + \chi^\dagger\vec{\sigma}\chi$$

Now in the non-relativistic limit ($pc \ll mc^2$ in $E^2 = p^2c^2 + m^2c^4$)

$$\chi = \frac{c\vec{\sigma}.\vec{p}}{E + mc^2}\phi \longrightarrow \frac{\vec{\sigma}.\vec{p}}{2mc}\phi = \frac{\vec{\sigma}.\hat{p}}{2}\frac{v}{c}\phi$$

so χ is smaller than ϕ by a factor v/c: we therefore neglect χ when it occurs to greater than first-order. We therefore have

$$\bar{\psi}\gamma^0\psi = \phi^\dagger\phi$$

$$\bar{\psi}\vec{\gamma}\psi = \frac{1}{2mc}\left\{\phi^\dagger\vec{\sigma}(\vec{\sigma}.\vec{p}\,\phi) + (\vec{\sigma}.\vec{p}\phi)^\dagger\vec{\sigma}\phi\right\}$$

$$\bar{\psi}\gamma_0\gamma_5\psi = \frac{1}{2mc}\left\{\phi^\dagger(\vec{\sigma}.\vec{p}\phi) + (\vec{\sigma}.\vec{p}\phi)^\dagger\phi\right\}$$

$$\bar{\psi}\vec{\gamma}\gamma_5\psi = \phi^\dagger\vec{\sigma}\phi$$

The parity-violating energy density can be written

$$E_{pv}(\vec{r}) = -\frac{G_F}{2\sqrt{2}} \sum_i \sum_b \{ \bar{e}\gamma_0(i)\gamma_5(i)e\,\bar{n}\gamma^0(b)n - \bar{e}\vec{\gamma}(i)\gamma_5(i)e\,\bar{n}\vec{\gamma}(b)n \}$$

Now $\bar{n}\vec{\gamma}(b)n$ contains terms in $\vec{p}_n(b)/m_nc$, where m_n is the neutron mass, while $\bar{e}\gamma_0(i)\gamma_5(i)e$ contains terms in $\vec{p}_e(i)/m_ec$, where m_e is the electron mass. We can neglect the neutron momentum (compared to m_nc - the nuclei are clearly not moving at relativistic velocities) but not the electron momentum (electrons in molecules do move at near-relativistic velocities). Thus, we can in fact neglect the "space" components in $j_\mu(e)j^\mu(n)$ leaving only the term from $j_0(e)j^0(n)$. We therefore have

$$E_{pv}(\vec{r}) = -\frac{G_F}{2\sqrt{2}} \sum_i \sum_b \frac{1}{2m_ec} \left\{ \begin{array}{l} \phi_e^\dagger(i)(\vec{\sigma}_e\cdot\vec{p}_e\phi_e(i)) \\ +(\vec{\sigma}_e\cdot\vec{p}_e\phi_e(i))^\dagger\phi_e(i) \end{array} \right\} \times \phi_n^\dagger(b)\phi_n(b)$$

for the parity-violating energy density at a point $\vec{r}$. Note that at this stage the γ-matrices have now been eliminated, so all authors should agree on the sign of this expression, regardless of which convention they used to establish the sign of $\mathcal{H}^{PV}(\vec{r})$.

To get the total PVES, this must be integrated over all space:

$$E_{pv} = \int E_{pv}(\vec{r})\,d^3\vec{r}$$

This can be expressed in terms of a Hamiltonian operator $\hat{H}_{pv}$:

$$E_{pv} = \sum_i \sum_b \iint \phi_e^\dagger(i)\phi_n^\dagger(b)\,\hat{H}_{pv}\,\phi_e(i)\phi_n(b)\,d^3\vec{r}_i\,d^3\vec{r}_b$$

To make these two equations compatible we write the first one as

$$E_{pv} = \sum_i \sum_b \iint \delta^3(\vec{r}_i - \vec{r}_b)\, E_{pv}(\vec{r}_i, \vec{r}_b)\, d^3\vec{r}_i\, d^3\vec{r}_b$$

where the delta-function expresses the contact nature of the weak interaction (Tranter 1984) and $E_{pv}(\vec{r}_i, \vec{r}_b)$ is now written as a function of the electron and neutron positions $\vec{r}_i$ and $\vec{r}_b$. To extract the Hamiltonian operator we must therefore equate

$$\phi_e^{+}(i)\phi_n^{+}(b)\, \hat{H}_{pv}\, \phi_e(i)\phi_n(b)$$

with

$$\delta^3(\vec{r}_i - \vec{r}_b)\, E_{pv}(\vec{r}_i, \vec{r}_b)$$

where

$$E_{pv}(\vec{r}_i, \vec{r}_b) = \frac{-G_F}{2\sqrt{2}}\left(\frac{1}{2m_e c}\right)\begin{Bmatrix} \phi_e^{+}(i)(\vec{\sigma}_e \cdot \vec{p}_e \phi_e(i)) \\ + (\vec{\sigma}_e \cdot \vec{p}_e \phi_e(i))^{\dagger} \phi_e(i) \end{Bmatrix}$$
$$\times\ \phi_n^{+}(b)\phi_n(b)$$

By realising that $(\vec{\sigma} . \vec{p}\, \phi)^{\dagger} = \phi^{\dagger} \vec{p}^{\dagger} \vec{\sigma}^{\dagger}$

it is readily shown that

$$\hat{H}_{pv} = \frac{-G_F}{2\sqrt{2}}\left(\frac{1}{2m_e c}\right)\sum_i \sum_b \begin{Bmatrix} \vec{\sigma}_e^{\dagger} . \vec{p}_e^{\dagger}\, \delta^3(\vec{r}_i - \vec{r}_b) \\ + \delta^3(\vec{r}_i - \vec{r}_b)\vec{\sigma}_e \cdot \vec{p}_e \end{Bmatrix}$$

or

$$\hat{H}_{pv} = -\frac{\Gamma}{2} \sum_i \sum_b \left\{ \vec{\sigma}(i) . \vec{p}(i),\ \delta^3(\vec{r}_i - \vec{r}_b) \right\}_{+}$$

where $\Gamma = \frac{G_F}{2\sqrt{2}} \frac{1}{2m_e c}$ -5.732×10^{-17} a.u. (Tranter 1984).

Now to get the PVES we take the expectation value over the ground state electronic and neutron wave functions:

$$E_{pv} = \int \cdots \int \psi_e^\dagger \psi_n^\dagger \hat{H}_{pv} \psi_e \psi_n \prod_i \prod_b d^3\vec{r}_i \, d^3\vec{r}_b$$

Now the only part of $\hat{H}_{pv}$ that depends on the neutron coordinates is

$$\delta^3(\vec{r}_i - \vec{r}_b)$$

Also

$$\int \cdots \int \psi_n^\dagger \left(\sum_b \delta^3(\vec{r}_i - \vec{r}_b) \right) \psi_n \prod_b d^3\vec{r}_b = \rho_n(\vec{r})$$

where $\rho_n(\vec{r}_i)$ is just the neutron density at the point $\vec{r}_i$. So now

$$\hat{H}_{pv} = (-\Gamma/2) \sum_i \{ \vec{\sigma}(i) \cdot \vec{p}(i), \rho_n(\vec{r}_i) \}_+$$

and E_{pv} is given by the expectation value of this over the electronic wave function only: the neutron wavefunctions have now gone. Since the nucleus is point-like, the neutron density can be written

$$\rho_n(\vec{r}_i) = \sum_a N_a \delta^3(\vec{r}_i - \vec{r}_a)$$

where N_a is the neutron number of nucleus a and the summation is over all nuclei a. Noting also that $\frac{1}{2}\vec{\sigma}(i) = \vec{s}(i)$, we obtain

$$\hat{H}_{pv} = -\Gamma \sum_i \sum_a N_a \{ \vec{s}(i) \cdot \vec{p}(i), \delta^3(\vec{r}_i - \vec{r}_a) \}_+$$

for our parity-violating Hamiltonian, with the sums over i and a being over all electrons and nuclei respectively.

This beautifully elegant expression summarizes the physical origin of the PVES. The term in $\vec{s}.\vec{p}$ represents the projection of the spin onto the direction of momentum, thus touching directly on the left-handedness of the electron. $\hat{H}_{pv}$ is of opposite sign for enantiomers: $\vec{p}$ changes sign under parity, being a polar vector, while $\vec{s}$ remains the same, being an angular momentum and therefore an axial vector. The delta function expresses the contact nature of the weak interaction: electrons only feel it when they are at the nucleus. The smallness of the PVES arises from the smallness of Γ, which contains the weak coupling constant G_F.

Getting the PVES as matrix elements of the parity-violating Hamiltonian over electronic wavefunctions

It would seem natural to separate the electronic wavefunctions Ψ into spin and orbital parts, $\Psi = \Psi_s \Psi_\ell$. We would then have

$$E_{pv} = -\Gamma \sum_i \sum_a N_a \langle \Psi_\ell | \{\vec{p}(i), \delta^3(\vec{r}_i - \vec{r}_a)\}_+ | \Psi_\ell \rangle \langle \Psi_s | \vec{s}(i) | \Psi_s \rangle$$

The spin matrix elements are always real or zero, but the orbital matrix elements are always imaginary or zero since $\vec{p} = (\hbar/i)\vec{\nabla}$. So E_{pv} is imaginary - but since as an observable it must be real ($\hat{H}_{pv}$ is a Hermitian operator) it can only have a value of zero.

The vanishing of E_{pv} is due to the assumed separability of Ψ into $\Psi_s \Psi_\ell$: but this separation is only possible in the absence of spin-orbit coupling (which mixes Ψ_s and Ψ_ℓ so that they are no longer independent). If the ground state wave function is corrected for the effect of spin-orbit coupling in mixing in excited states Ψ_f it becomes

$$|\Psi_o'\rangle = |\Psi_o\rangle + \sum_{f \neq 0} \frac{\langle \Psi_f | \hat{H}_{so} | \Psi_o \rangle}{E_o - E_f} |\Psi_f\rangle$$

We therefore obtain

$$E_{pv} = \langle \Psi_0' | \hat{H}_{pv} | \Psi_0' \rangle$$

$$= 2 \sum_{f \neq 0} \mathrm{Re} \left\{ \frac{\langle \Psi_0 | \hat{H}_{pv} | \Psi_f \rangle \langle \Psi_f | \hat{H}_{so} | \Psi_0 \rangle}{E_0 - E_f} \right\}$$

for the ground-state PVES, where

$$\hat{H}_{so} = \sum_j \sum_b \xi(b,j) \vec{\ell}(b,j) \cdot \vec{s}(j)$$

with $\xi(b,j)$ the spin-orbit coupling constant for electron j in the field of nucleus b.

We currently have many-electron wavefunctions Ψ and Hamiltonians $\hat{H}_{pv}$ and $\hat{H}_{so}$. The latter can be written in terms of single electron operators $\hat{h}_{pv}$ and $\hat{h}_{so}$

$$\hat{H}_{pv} = \sum_i \hat{h}_{pv}(i), \quad \hat{H}_{so} = \sum_j \hat{h}_{so}(j),$$

$$\hat{h}_{pv}(i) = -\Gamma \sum_a N_a \{ \vec{s} \cdot \vec{p}, \delta^3(\vec{r}_i - \vec{r}_a) \}_+$$

$$\hat{h}_{so}(i) = \sum_b \xi(b) \vec{\ell}(b,j) \cdot \vec{s}(j)$$

and the total molecular electronic wavefunction Ψ is written as a product of single electron spin-orbital wave functions ψ,

$$\Psi_0 = \psi_j(1) \psi_{j'}(2) \psi_{j''}(3) \cdots\cdots$$

where $\psi_j(1)$ represents electron 1 in spin-orbital j, etc.. We consider only ground states Ψ_0 in which all electrons are spin-paired, and only singly-excited excited states Ψ_f, e.g.

$$\Psi_0 = \psi_j(1)\,\psi_{j'}(2)\,\psi_{j''}(3)\,\psi_{j'''}(4)$$

$$\Psi_f = \psi_j(1)\,\psi_{j'}(2)\,\psi_k(3)\,\psi_{j'''}(4)$$

Here the electron that was in $\psi_{j''}$ has been excited to an unoccupied orbital ψ_k so we therefore have

$$\langle \Psi_f | \hat{H} | \Psi_0 \rangle = \langle \psi_k | \hat{h} | \psi_{j''} \rangle$$

In the expression for E_{pv}, we sum over all excited states Ψ_f, which means considering electrons in all possible occupied orbitals ψ_j being excited to all possible unoccupied orbitals ψ_k, i.e.

$$E_{pv} = 2 \sum_j^o \sum_k^u \mathrm{Re} \left\{ \frac{\langle \psi_j | \hat{h}_{pv} | \psi_k \rangle \langle \psi_k | \hat{h}_{so} | \psi_j \rangle}{\epsilon_j - \epsilon_k} \right\}$$

where ϵ_j, ϵ_k are the energies of the individual 1-electron wavefunctions ψ, $\sum_j^o$ is the sum over all MOs occupied in the ground state, and $\sum_k^u$ is the sum over all MOs unoccupied in the ground state. We can write

$$E_{pv} = \sum_j^o \sum_k^u P_{jk} / (\epsilon_j - \epsilon_k)$$

where

$$P_{jk} = 2\,\mathrm{Re} \langle \psi_j | \hat{h}_{pv} | \psi_k \rangle \langle \psi_k | \hat{h}_{so} | \psi_j \rangle$$

is the "parity-violating strength" for a virtual transition $j \rightarrow k$ between molecular spin-orbitals.

Next we separate the molecular spin-orbitals into spin and orbital parts,

$$|\psi_j\rangle = |\psi_j^{\ell}\rangle |\psi_j^{s}\rangle$$

so that we therefore have

$$E_{pv} = 2\sum_j^{o}\sum_k^{u} \mathrm{Re}\frac{\left\{\langle\psi_j^{\ell}|\hat{V}_{pv}|\psi_k^{\ell}\rangle\cdot\langle\psi_j^{s}|\vec{s}|\psi_k^{s}\rangle \times \langle\psi_k^{\ell}|\hat{V}_{so}|\psi_j^{\ell}\rangle\cdot\langle\psi_k^{s}|\vec{s}|\psi_j^{s}\rangle\right\}}{(\epsilon_j - \epsilon_k)}$$

where $$\hat{V}_{pv} = -\Gamma\sum_a N_a\{\vec{p}, \delta^3(\vec{r}-\vec{r}_a)\}_+$$

$$\hat{V}_{so} = \sum_b \xi(b)\vec{\ell}(b)$$

(these operators are just $\hat{h}_{pv}$, $\hat{h}_{so}$ without the $\vec{s}$ dotted on).

The sums over j and k are currently over molecular spin-orbitals (in the unrestricted Hartree-Fock, the orbital part may be different for spin-up and spin-down): $|\psi_j^{\ell}\rangle|+m_s\rangle$ and $|\psi_j^{\ell}\rangle|-m_s\rangle$ are regarded as different spin-orbitals within in the summation. If we move to the restricted Hartree-Fock in which the orbital part is independent of spin orintation, the sum $\sum_j$ over molecular spin-robitals is replaced by $\sum_j\sum_{m_s}$, in which the summation over j now refers to spin-free MOs. To effect the spin summation, we write $\langle\psi_j^{\ell}|\hat{V}_{pv}|\psi_k^{\ell}\rangle$ as the vector $\vec{A}$ and $\langle\psi_k^{\ell}|\hat{V}_{so}|\psi_j^{\ell}\rangle$ as the vector $\vec{B}$, and we need to evaluate

$$\sum_{m_s'}\sum_{m_s} \mathrm{Re}\left\{A_x\langle m_s'|\hat{s}_x|m_s\rangle + A_y\langle m_s'|\hat{s}_y|m_s\rangle + A_z\langle m_s'|\hat{s}_z|m_s\rangle\right\}$$
$$\times\left\{B_x\langle m_s|\hat{s}_x|m_s'\rangle + B_y\langle m_s|\hat{s}_y|m_s'\rangle + B_z\langle m_s|\hat{s}_z|m_s'\rangle\right\}$$

From the familiar properties of angular momentum operators (Atkins 1983), the only non-zero matrix elements are (putting $\hbar = 1$, as we work in atomic units):

$$\langle\alpha|\hat{s}_z|\alpha\rangle = +1/2, \quad \langle\beta|\hat{s}_z|\beta\rangle = -1/2$$

$$\langle\alpha|\hat{s}_x|\beta\rangle = 1/2, \quad \langle\beta|\hat{s}_x|\alpha\rangle = 1/2$$

$$\langle\alpha|\hat{s}_y|\beta\rangle = \frac{1}{2i}, \quad \langle\beta|\hat{s}_y|\alpha\rangle = -1/2i$$

The contributions to the required expression for the 4 possible spin contributions are

$$m_s = m_s' = \alpha : \quad 1/4\, A_z B_z$$
$$m_s = m_s' = \beta : \quad 1/4\, A_z B_z$$
$$m_s = \alpha,\ m_s' = \beta : \quad 1/4\, A_x B_x + 1/4\, A_y B_y$$
$$m_s = \beta,\ m_s' = \alpha : \quad 1/4\, A_x B_x + 1/4\, A_y B_y$$

This makes a total of $(1/2)\vec{A}.\vec{B}$, so the PVES now becomes

$$E_{pv} = \mathrm{Re} \sum_j^{o} \sum_k^{u} \frac{\langle\psi_j|\hat{V}_{pv}|\psi_k\rangle \cdot \langle\psi_k|\hat{V}_{so}|\psi_j\rangle}{\epsilon_j - \epsilon_k}$$

where ψ_j are now the spin-free MOs and

$$\bar{P}_{jk} = \mathrm{Re}\langle\psi_j|\hat{V}_{pv}|\psi_k\rangle \cdot \langle\psi_k|\hat{V}_{so}|\psi_j\rangle$$

is the spin-independent parity-violating strength for the transition $j \rightarrow k$. Note the similarity between $\bar{P}_{jk}$ and the rotational strength

$$R_{jk} = \mathrm{Im}\langle\psi_j|\hat{\mu}|\psi_k\rangle \cdot \langle\psi_k|\hat{m}|\psi_j\rangle$$

in optical activity: $\hat{\mu}$ and $\hat{V}_{pv}$ are polar vectors (change sign under parity) while $\hat{m}$ and $\hat{V}_{so}$ are axial vectors (unchanged by parity), so overall both $\bar{P}_{jk}$ and R_{jk} are pseudoscalars (opposite sign for enantiomers).

Evaluating the PVES in the LCAO approximation

In the LCAO approximation the MOs are expanded as a linear combination of atomic orbitals,

$$|\psi_j\rangle = \sum_c \sum_\gamma C^j_{c\gamma} |\phi_{c\gamma}\rangle$$

where the sums are over all nuclei c and over all types of AO, γ , on nucleus c. The PVES for a singlet ground state therefore becomes

$$E_{pv} = \sum_j^o \sum_k^u \sum_c \sum_{c'} \sum_{c''} \sum_{c'''} \sum_\gamma \sum_{\gamma'} \sum_{\gamma''} \sum_{\gamma'''} C^{j*}_{c\gamma} C^k_{c'\gamma'} C^{k*}_{c''\gamma''} C^j_{c'''\gamma'''}$$

$$\mathrm{Re}\left\{ \frac{\langle \phi_{c\gamma} | \hat{V}_{pv} | \phi_{c'\gamma'} \rangle \cdot \langle \phi_{c''\gamma''} | \hat{V}_{so} | \phi_{c'''\gamma'''} \rangle}{\epsilon_j - \epsilon_k} \right\}$$

where

$$\hat{V}_{pv} = -\Gamma \sum_a N_a \{ \vec{p}, \delta^3(\vec{r} - \vec{r}_a) \}_+$$

$$\hat{V}_{so} = \sum_b \xi(b) \vec{\ell}(b)$$

The operators $\hat{V}_{pv}$ and $\hat{V}_{so}$ are clearly most effective in the regions close to the nuclei a and b respectively, so it is reasonable to assume that

$$\langle \phi_{c\gamma} | \hat{V}_{pv} | \phi_{c'\gamma'} \rangle = 0 \quad \text{unless} \quad c = c' = a$$

$$\langle \phi_{c''\gamma''} | \hat{V}_{so} | \phi_{c'''\gamma'''} \rangle = 0 \quad \text{unless} \quad c'' = c''' = b$$

Our final expression for the PVES therefore becomes (Tranter 1984)

$$E_{pv} = -\Gamma \sum_j^{o} \sum_k^{u} \sum_a \sum_b \sum_\gamma \sum_{\gamma'} \sum_{\gamma''} \sum_{\gamma'''} C^{j*}_{a\gamma} C^{k}_{a\gamma'} C^{k*}_{b\gamma''} C^{j}_{b\gamma'''}$$

$$\times \frac{N_a \, \mathrm{Re} \left\{ \langle \phi_{a\gamma} | \{ \vec{p}, \delta^3(\vec{r} - \vec{r}_a) \}_+ | \phi_{a\gamma'} \rangle \cdot \langle \phi_{b\gamma''} | \xi(b) \vec{l}(b) | \phi_{b\gamma'''} \rangle \right\}}{\epsilon_j - \epsilon_k}$$

The coefficients $C^{k}_{a\gamma'}$ etc. of the AOs in each MO, along with the energy ϵ_k of each MO, are supplied for any given molecular geometry by the GAUSSIAN ab initio package using a STO-6-31G basis set. These are then fed into our own program PVED84, which takes the values of the matrix elements $\langle \phi_{a\gamma} | \hat{V}_{pv} | \phi_{a\gamma'} \rangle$ etc, and uses the coefficients $C^{k}_{a\gamma'}$ etc. to effect the required summation and finally print out the value of the parity-violating energy difference between enantiomers,

$$\Delta E_{pv} = 2 E_{pv}$$

(Tranter 1984, Mason & Tranter 1984, 1985).
Typical values of the PVED are about 10^{-20}a.u. or 10^{-17} kT for molecules containing first-row atoms. Note that there is nothing small about the matrix elements: the result only becomes small on multiplication by Γ, which contains the very small weak interaction constant G_F.

The final expression for E_{pv} has some interesting features. Firstly we note that each term is a product of a parity-violating matrix element on one atom a and a spin-orbit coupling matrix element on the same or another atom b. However there is no reference in the expression to the distance between atoms a and b: this means that in a large molecule, cross-terms between even very distant pairs of atoms have to be included - although they may not contribute much unless the corresponding MOs are delocalized to encompass both atom a and atom b. Secondly we notice that E_{pv} is proportional to Z^5: N_a is proportional to Z, ξ(r) to Z, ϕ to $Z^{3/2}$ (from the normalization constant in hydrogenic orbitals), and hydrogen-like orbital energies are proprortional to Z^2. This means that although typical

PVEDs for molecules containing first row atoms are only 10^{-17} kT, PVEDs for molecules containing heavier atoms are likely to be considerably larger.

We will not go into the details of the evaluation of the matrix elements of $\hat{V}_{pv}$ and $\hat{V}_{so}$ between the AOs ϕ (see Tranter 1984). Suffice it to say that $\hat{V}_{pv}$ only connects s and p orbitals (to be expected since they have opposite parity), while $\hat{V}_{so}$ only connects p orbitals (again expected since s orbitals have zero orbital angular momentum). We have hitherto neglected d orbitals in our calculations on first and second row atoms: in any case they never contribute to the matrix elements of $\hat{V}_{pv}$, although they do contribute to those of $\hat{V}_{so}$; but the denominator $\epsilon_j - \epsilon_k$ is probably prohibitively large for MOs involving the high energy d orbitals.

E. AMPLIFICATION MECHANISMS

Typical PVEDs are of the order of 10^{-17} kT. Under conditions of thermodynamic control this results, from the Boltzmann distribution, in an enantiomeric excess of $|L-D|/(L+D) = 10^{-17}$. Similar enantiomeric excesses are also obtained under conditions of kinetic control because the activation energies of enantiomers are unequal.

These small enantiomeric excesses need amplifying in order to result in the observed biohomochirality, and the proposed mechanisms fall into two classes: (a) time-extended, uniformly <u>cumulative</u> chiral amplification; and (b) <u>catastrophic</u> bifurcation by a chiral perturbation over a relatively short time.

The Yamagata cumulative mechanism (Yamagata 1966) applies to polymerizations or crystallizations involving labile monomer units, e.g. the formation of chiral quartz crystals from achiral silica units. An achiral unit A may add on to a growing crystal of either hand:

$$L_{n-1} + A \rightleftharpoons L_n$$

$$\Delta G_L \neq \Delta G_D$$

$$D_{n-1} + A \rightleftharpoons D_n$$

But owing to the PVED the corresponding free energy changes are not quite the same, which means that A will add preferentially to one hand of crystal rather than the other, resulting in a fractional excess

$$f_L = \frac{L_n/L_{n-1}}{D_n/D_{n-1}} = e^{-(\Delta G_L - \Delta G_D)/RT}$$

This fractional excess is operative at each of N stages of crystallization, and leads cumulatively to an excess of one hand of crystal given by

$$\frac{L_N}{D_N} = \frac{L_1}{D_1} f_L^{N-1} = (1 + (\Delta G_L - \Delta G_D)/RT)^N$$

$$= 1 + N(\Delta G_L - \Delta G_D)/RT$$

This is equivalent to the PVED of the crystal being N times the PVED of the individual units:

$$\Delta E_{pv}(\text{N-unit crystal}) = N\Delta E_{pv}(\text{one unit})$$

The Kondepudi catastrophic mechanism (Kondepudi & Nelson 1983, 1984, 1985, Kondepudi 1987) is based on a kinetic scheme involving autocatalysis and enantiomeric antagonism, i.e. the presence of one enantiomer encourages production of itself, but inhibits production of its enantiomer. Many polymerization reactions essential to life have precisely these characteristics, and some enantiomeric enrichment can be demonstrated in the laboratory (Lundberg & Doty 1957, Joyce et al 1984, Brack & Spach 1971, 1979, 1980). Whereas the Yamagata mechanism requires the monomers to be labile, the Kondepudi mechanism requires them to be non-labile, as in the polymerization of amino acids for example.

Kondepudi envisages an open-flow reactor system, such as a lake, fed by an input of achiral substances, A and B, with an output of enantiomers X_L and X_D and other products. The enantiomers form reversibly from the substrates, both directly (k_1) and autocatalytically (k_2), while cross-inhibition between the enantiomers results in their irreversible conversion (k_3) to products P:

$$A + B \underset{k_{-1}}{\overset{k_1}{\rightleftharpoons}} X_{L(D)}$$

$$X_{L(D)} + A + B \underset{k_{-2}}{\overset{k_2}{\rightleftharpoons}} 2X_{L(D)}$$

$$X_L + X_D \xrightarrow{k_3} P$$

The scheme can accommodate unequal reaction rates for the two enantiomers, and can be extended to include racemization, thermal fluctuations, etc.. With an input of A and B maintaining the substrate concentration at a constant or slowly increasing level, and a corresponding output of products, the system attains a dynamic quasi-steady state far removed from thermodynamic equilibrium. Using the methods of non-equilibrium statistical thermodynamics it can be shown that as the input concentration is increased, the system reaches a transition point, where the racemic production process becomes metastable and the symmetry is spontaneously broken, with bifurcation into homochiral production of one enantiomer or the other. Without a chiral influence, the choice of enantiomer is arbitrary, but at the transition point the system becomes hypersensitive to small chiral perturbations or fluctuations, which may cause the bifurcation to become determinate. It is found (Kondepudi & Nelson 1984, 1985) that thermal fluctuations damp out any initial trend toward one hand or the other unless $\Delta E/kT > 10^{-17}$, where ΔE may represent either the difference in activation energies between enantiomers (Kondepudi & Nelson 1985, 1985) or the PVED between enantiomers in an otherwise racemic substrate input (Hegstrom 1985). Kondepudi (1987) estimates that it takes about 10^4 years to amplify to homochirality from the enantiomeric excess of 10^{-17} associated with the PVED. This figure is based on realistic production rates of 10^{-10}mol $L^{-1}s^{-1}$, and modest concentrations rising from 3×10^{-3}M to 6×10^{-3}M in a flow system the size of a small lake of volume 4×10^9L (1kmx1kmx4m). Amplification can be achieved in the same time with lower reaction rates and concentrations if the lake is larger.

Racemization obviously opposes any amplification effect, which is why the reactants are supposed to be non-labile in the Kondepudi mechanism. However, if the reaction scheme is extended to include a racemization step, it is found that amplification in 10^4 years from a 10^{-17} excess under the above conditions can withstand a racemization half-life as low as 10^2-10^3

years (Kondepudi 1987), compared with typical values of 10^5-10^6 years for most amino acids (Bada & Miller 1987).

Although typical PVEDs of 10^{-17}kT are only just large enough to be amplified, the amplification time is a very sensitive function of the PVED. An increase in the PVED of one order of magnitude decreases the amplification time by four orders of magnitude. Thus, whereas a PVED of 10^{-17} kT takes 10^4 years to amplify, a PVED of 10^{-16} kT takes only one year to amplify. These figures present a most tantalizing prospect for electroweak bioenantioselection: it is either very easy or only just possible, depending on the precise value of the PVED within one order of magnitude.

The lake volume is also important: an increase in the PVED of one order of magnitude allows the lake volume to decrease by two orders of magnitude for the same amplification time. The amplification time is less sensitive to concentrations, reaction rates, etc.: if the concentration increases by one order of magnitude, the amplification time decreases by just one order of magnitude.

Thus, in summary, the Kondepudi mechanism allows amplification from the PVED enantiomeric excess of 10^{-17} in 10^4 years and $4x10^9$L, which is reasonable on an evolutionary scale. The enantiomeric excess of 10^{-12} theoretically obtainable from β-radiolysis could, however be amplified in just 1 year and 40L, which could be verified on a laboratory scale.

F. PVEDS OF SOME IMPORTANT BIOMOLECULES

Mason and Tranter have calculated the PVED of some α-amino acids in the zwitterionic form $\overset{+}{N}H_3CHRC\overset{-}{O_2}$ in the preferred solution conformation (Mason & Tranter 1983, 1984, 1985). L-alanine, L-valine, L-serine and the L-aspartate anion are all found to be stabilized relative to their D-enantiomers by about 10^{-17} kT, just enough for Kondepudi amplification to homochirality. The total PVES of a molecule can be resolved into contributions from each nucleus. Hydrogen nuclei give a zero contribution, as they have no neutrons, but their presence does influence the PVES of the other nuclei through its effect on the molecular orbitals. For the amino acids studied, the major contributions are from the C_α chiral centre and the electron-rich carboxylate oxygens, and these are consistently negative in all cases. The contribution from the side group R is consistently positive, and increases in magnitude on going from alanine through valine and serine to the aspartate anion, following the approximate order of increasing side-group polarizability. This is because the side groups are achiral in themselves, and are able to contribute only if they acquire a degree of chirality by the rest of the molecule asymmetrically polarizing them. The relatively small positive contribution from the side group does not, however, prevent the overall PVES being negative for all L-amino acids studied. It is therefore likely that this electroweak stabilization of the natural L enantiomers is a general result for all L-α-amino acids with achiral side-groups of limited polarizability (Tranter 1985a,b). Furthermore the L-amino acids were found to be PVED-stabilized not only in the solution conformation,but also in the α-helix and β-sheet conformations (Mason & Tranter 1984, 1985). Enantiomeric differences in activation energies were also found to favour the L-amino acids in a prebiotic reaction path (Tranter 1985c, 1986a,b,c).

Tranter and MacDermott are now examining whether this success with the L-amino acids can be repeated for the D-sugars. The aldoses, $(H_2CO)_n$, were investigated for n = 3 to 5, but the results are not so clear-cut as for the amino acids (Tranter & MacDermott et al in preparation). For glyceraldehyde (n=3), the parent of the higher sugars, the D form is PVED-stabilized by about 10^{-17} kT (Tranter 1986c); D-deoxyribose is also PVED-stabilized in the C2-endo form found in DNA; but D-ribose in the C3-endo form found in RNA is <u>less</u> stable than its enantiomer (Tranter & MacDermott et al in preparation). The latter results are explained by the PVED of the basic furanose sugar skeleton, tetrahydrofuran (THF): the C2-endo ring conformation of THF is more stable than the enantiomeric C3-endo conformation by about 10^{-17} kT (Tranter & MacDermott 1986).

Although D-ribose cannot have been selected by its own PVED (since L-ribose is more stable), the D-sugar series may have been selected by the PVED of its parent D-glyceraldehyde, or alternatively by the diastereomeric connection with the PVED-stabilized L-amino acids. However it is very probable that ribose, being difficult to synthesize, is not prebiotic but represents a later biological evolution after the first self-replicating systems appeared, in which case its own PVED is irrelevant. It is therefore necessary to consider more primitive replicators than DNA or RNA: for example, duplex formation with base-pairing can occur with a glycerol rather than ribose-based polymer (Bada & Miller 1987, Schwartz & Orgel 1985) in which the monomer units are achiral.

Thus, both amino acids and sugars have PVEDs of the order of 10^{-17} kT. In the search for slightly larger PVEDs - bearing in mind that 10^{-17} kT is only just amplifiable and takes 10^4 years, while 10^{-16} kT is easily amplified in just one year - MacDermott and Tranter (1989) turned to molecules containing heavier atoms, since the PVED is proportional to the fifth power of the atomic number. They considered (MacDermott & Tranter in preparation) deoxyribose-type rings without the 1-hydroxyl group but with a phosphate group at either the 3′ position or the 5′ position. These molecules represent the building blocks of the sugar-phosphate backbone of DNA, but without the bases. It had been hoped that these molecules would have larger PVEDs due to the heavy phosphorus atom, but the PVEDs were again only of the order of 10^{-17} kT: this was found to be partly because the P atoms are quite electropositive in phosphates, so there is not enough electron density to feel the potentially larger effect.

The contributions to the PVES from the 3′ and 5′ phosphate groups were of opposite sign; to assess the PVES of the DNA backbone it is therefore necessary to consider a phosphate group between two deoxyribose-type rings, attached to the 3′ position of one and the 5′ position of the other. A C2-endo sugar ring between two phosphates (one 3′ and one 5′) was found to be PVED-stabilized, with a PVES of similar magnitude to that of the ring in deoxyribose or THF. But the phosphate between two sugar rings forms a molecule that is too large for GAUSSIAN82. Therefore, to assess the contribution of the phosphate group to the PVED of DNA-analogues, a "minimal" helical backbone, with all ring atoms removed except the essential C5,C4 and C3, was considered (MacDermott & Tranter in preparation), using B-DNA torsion angles. In this system the PVES of the phosphate group was negative, resulting in the right-hand helical backbone being stabilized by about 10^{-17} kT per monomer unit.

This raises the interesting question of whether the D-sugars fixed or followed the right-hand double helix. It is usually assumed that the D-sugars fixed the right-handedness of the DNA helix by diastereomeric preference. But it could be that the right-hand helix was selected in some earlier, simpler replicator (perhaps something resembling the "minimal" helical backbone or the glycerol based double helix), in which case steric considerations in any later ring-closure to give ribose would direct selection of the D-sugars.

Another important system which contains both helices and second-row atoms is quartz: the l(-)-form consists of right-hand 3-fold and left-hand 6-fold helices of silica tetrahedra. As a chiral mineral made of achiral units it can undergo cumulative chiral amplification by the Yamagata mechanism, and indeed a 1.4% excess of l(-)-quartz has been reported in a total collection of 16 807 crystals from all over the world (Vistelius 1950, Palache et al 1962). MacDermott and Tranter (ms in press and ms submitted) have recently calculated the PVES of Si and O atoms at the centre of small fragments of quartz (atoms at the edges will obviously give atypical PVES) and found, most gratifyingly, that l(-)-quartz is indeed PVED-stabilized, by about 10^{-17} kT per SiO_2 unit. This is again disappointingly small for a second-row atom, and this is again because the Si is highly electropositive in this system. However, the PVED does not need to be any larger than 10^{-17} kT to account for the 1.4% excess of l(-)-quartz. According to the Yamagata mechanism,

$$\Delta E_{pv}(\text{crystal}) = N \Delta E_{pv}(\text{unit})$$

so ΔE_{pv}(crystal) = 10^{-2} kT (corresponding to a 1% enantiomeric excess) can be obtained from ΔE(unit) = 10^{-17} kT if N = 10^{15}, which corresponds (Tranter 1985c) to a realistic crystal of side 0.1mm.

These new PVED results for quartz are therefore tremendously exciting because they predict almost exactly the observed 1% excess of l(-)-quartz, so that this excess can now for the first time be regarded as evidence for the global symmetry-breaking effects of the weak interaction. Furthermore, catalysis by quartz-like minerals presents an opportunity for their chiral bias to be transferred to biology, for l(-)-quartz has been shown to adsorb L-alanine preferentially from a racemic mixture, for purely diastereomeric reasons, with an enantioselectivity of 1% or more (Kavasmaneck & Bonner 1977, Furuyama et al 1982). Combining this asymmetry of 10^{-2} in adsorption with the 10^{-2} asymmetry in the quartz crystals themselves gives (Tranter 1985c) an overall electroweak enantioselectivity of 10^{-4}, which is much greater than that of 10^{-17} from the PVED of individual molecules. Such a large enantiomeric excess would require only minor subsequent amplification, and homochirality could be achieved in relatively short times using small reaction volumes. Realistic chiral mineral catalysts are not quartz itself, but aluminosilicate quartz analogues, possibly containing heavy metal cations which might produce larger PVEDs.

In conclusion, the weak interaction appears to be the only consistent universal chiral influence. Most calculations of the PVED carried out so far do seem to favour the natural enantiomers: L-amino acids, D-glyceraldehyde, D-deoxyribose, right-hand "minimal" helices, and l(-)-quartz (containing right-hand 3-fold helices) are all PVED-stabilized. Future studies - in search of larger PVEDs - will be directed towards larger systems, especially polymers and crystals, containing electronegative heavy atoms.

REFERENCES

Aitchison, I.J.R., and Hey, A.J.G., 1982, Gauge Theories in Particle Physics, Adam Hilger, Bristol.
Angel, J.R.P., Illing, R., and Martin, P.G., 1972, Nature, 238:389.
Atkins, P.W., 1983, Molecular Quantum Mechanics, Oxford.
Atwater, H.A., 1974, Introduction to General Relativity, Pergamon.
Bada, J.L., and Miller, S.L., 1987, BioSystems, 20:21.
Barra, A.L., Robert, J.B., and Wiesenfeld, L., 1987, BioSystems, 20:57.
Barron, L.D., 1981, Mol.Phys., 43:1395.
Barron, L.D., 1986a, Chem.Soc.Rev., 15:189.
Barron, L.D., 1986b, Chem.Phys.Lett., 123:423.
Barron, L.D., 1987a, BioSystems, 20:7.
Barron, L.D., 1987b, Chem.Phys.Lett. 135:1.
Binkley, J.S., Frisch, M.J., Raghavachari, K., DeFrees, D., Schlegel, H.B., Whiteside, R.A., Fluder, E., Seeger, R., and Pople, J.A., GAUSSIAN82, Carnegie-Mellon University, Pittsburgh.
Birss, R.R, 1966, Symmetry and Magnetism, North Holland, Amsterdam.
Blair, N.E., Dirbas, F.M., and Bonner, W.A., 1981, Tetrahedron, 37:27.
Bonner, W.A., 1984, Orig.Life, 14:383.
Borchkhade, T.M., and Kogoshvili, N.G., 1976, Astron.Astrophys., 53:431.
Bouchiat, C.C., and Bouchiat, M.A., 1974a, J.Phys.(Paris), 35:899.
Bouchiat, M.A., and Bouchiat, C.C., 1974b, Phys.Lett. B, 48:111.
Bouchiat, M.A., and Bouchiat, C.C., 1975, J.Phys.(Paris), 36:493.
Brack, A., and Spach, G., 1971, Nature ps, 229:124.
Brack, A., and Spach, G., 1979, J.Mol.Evol., 13:35,47.
Brack, A., and Spach, G., 1980, J.Mol.Evol., 15:231.
Calder, N., 1977, The Key to the Universe, BBC.
Campbell, D.M., and Farago, P.S., 1985, Nature, 318:52.
Cairns-Smith, A.G., 1986, Chem.Brit., 22:559.
Dirac, P.A.M., Principles of Quantum Mechanics, Oxford 1958.
Dougherty, R.C., 1980, J.Am.Chem.Soc., 102:380.
Dougherty, R.C., 1981, Origins Life, 11:71.
Edwards, D., Cooper, K., and Dougherty, R.C., 1980,J.Am.Chem.Soc., 102:381.
Emmons, T.P., Reeves, J.M., and Fortson, E.N., 1983, Phys.Rev.Lett., 51:2089; 1984, ibid, 52:86.
Fischer, E., 1894, Chem. Ber., 27:2985, 3189.
Fortson, E.N., and Wilets, L., 1980, Adv.Atom.Molec.Phys., 16:319.
Furuyama, S., Sawada, M., Machiya, K., and Morimoto, T., 1982, Bull.Chem.Soc.Jap., 55:3394.
Gerike, P., 1975, Naturwissenschaften, 62:38.
Hegstrom, R.A., 1985, Nature, 315:749.
Hegstrom, R.A., 1987, BioSystems, 20:49.

Hegstrom, R.A., Rein, D.W., and Sandars, P.G.H., 1979, Phys.Lett A, 71:499.
Hegstrom, R.A., Rein,D.W., and Sandars, P.G.H., 1980, J.Chem.Phys., 73:2329.
Hegstrom, R.A., Rich, A., and Van House, J., 1985, Nature, 313:391.
Huang, K., 1982, Quarks, Leptons & Gauge Fields, World Scientific.
Idelson, M., and Blout, E.R., 1958 J.Am.Chem.Soc., 80:2387.
Izumi, Y., and Tai, A., 1977, Stereo-differentiating Reactions, Academic Press, New York.
Joyce, G.F., Visser, G.M., van Boeckel, C.A.A., van Boom, J.H., Orgel, L.E., and van Westresen, J., 1984, Nature, 310:602.
Kavasmaneck, P.R., and Bonner, W.A., 1977, J.Am.Chem.Soc., 99:44.
Kondepudi, D.K., 1987, BioSystems, 20:75.
Kondepudi, D.K., and Nelson, G.W., 1983, Phys.Rev.Lett., 50:1023.
Kondepudi, D.K., and Nelson, G.W., 1984, Physica A, 125:465.
Kondepudi, D.K., and Nelson, G.W., 1985, Nature, 314:438.
Letokhov, V.S., 1975, Phys.Lett. A, 53:275.
Lundberg, R.D., and Doty, P.J., 1957, J.Am.Chem.Soc., 79:3961.
MacDermott, A.J., and Tranter, G.E., 1989, Croatica Chemica Acta 62:165.
MacDermott, A.J., and Tranter, G.E., Chem.Phys.Lett. in press.
MacDermott, A.J., and Tranter, G.E., ms submitted to Nature.
MacDermott, A.J., and Tranter, G.E., ms in preparation, to be submitted to Proc.R.Soc.A Lond.
MacGillivray, H.T., and Dodd, R.J., 1985, Astron.Astrophys., 145:269.
Mason, S.F., 1982, Molecular Optical Activity and the Chiral Discriminations, Cambridge University Press.
Mason, S.F., 1983, Int.Rev.Phys.Chem., 3:217.
Mason, S.F., and Tranter, G.E., 1983, J.Chem.Soc.Chem.Commun., p.117.
Mason, S.F., and Tranter, G.E., 1984, Mol.Phys., 53:1091.
Mason, S.F., and Tranter, G.E., 1985, Proc.R.Soc.Lond.A, 397:45.
Mead, C.A., and Moscowitz, A., 1980, J.Am.Chem.Soc., 102:7301.
Melcher, G.,1974, J.Mol.Evol., 3:121.
Miller, S.L., and Orgel, L.E., 1974, The origins of life on the Earth, Prentice Hall, Englewood Cliffs, p.171.
Palache, C., Berman, H., and Frondel, C., 1962, Dana's System of Mineralogy, 7th ed. Vol.III, Wiley, New York, p.16.
Pasteur, L., 1848, Compt.Rend.Seances Soc.Biol.Paris, 26:535.
Pasteur, L., 1874, Compt.Rend.Acad.Sci.(Paris), 78:1515.
Pasteur, L., 1884a, Rev.Scientifique, 7:2.
Pasteur, L., 1884b, Bull.Soc.Chim.France, 41:215.
Pasteur, L., 1922, Oeuvres de Pasteur, Vol I, Dissymetrie Moleculaire, ed. Pasteur Vallery-Radot, Masson et Cie, Paris, p.369.
Rein, D.W., 1974, J.Mol.Evol., 4:15.
Sakurai, J.J., 1976, Advanced Quantum Mechanics, Addison-Wesley.
Salam, A., 1968, in Proc 8th Nobel Symp., Elementary particle theory (ed.N.Svartholm), Almquist & Wiksell, Stockholm.
Sandars, P.G.H., 1980, Fundamental interactions and structure of matter (ed. K.Crowse, J.Duclos, G.Fiorentini and G.Torelli), Plenum, New York, p.57.
Schwartz, A.W., and Orgel, L.E., 1985, Science, 228:585.
Spach, G., 1974, Chimia, 28:500.
Tranter, G.E., 1984, Ph.D. thesis, King's College, University of London.
Tranter, G.E., 1985a, Chem.Phys.Lett., 120:93.
Tranter, G.E., 1985b, Mol.Phys., 56:825.
Tranter, G.E., 1985c, Chem.Phys.Lett., 115:286.
Tranter, G.E., 1985d, Nature, 318:172.
Tranter, G.E., 1986a, J.Theor.Biol., 119:467.
Tranter, G.E., 1986b, BioSystems, 20:37.
Tranter, G.E., 1986c, J.Chem.Soc.Chem.Commun., p.60.
Tranter, G.E., 1987, Chem.Phys.Lett., 135:279.

Tranter, G.E., and MacDermott, A.J., 1986, Chem.Phys.Lett., 130:120.
Tranter, G.E., MacDermott, A.J., Overill, R.E., and Spears, P.J., ms in preparation, to be submitted to Proc.R.Soc.A Lond.
Van House, J., Rich, A., and Zitzewitz, P.W., 1984, Orig. Life, 14:413.
Vistelius, A.B., 1950, Zapiski Vsyesoyuz.Mineral.Obsh., 79:191.
Walgate, R., 1983, Nature, 303:473.
Weinberg, S., 1967, Phys.Rev.Lett., 19:1264.
Wolfrom, M.L., Lemieux, R.U., and Olin, S.M., 1949, J.Am.Chem.Soc., 71:2870.
Wolstencroft, R.D., 1984, IAU Symposium 112, Boston.
Wu, C.S., Ambler, E., Hayward, R.W., Hoppes, D.D., and Hudson, R.P., 1957, Phys.Rev., 105:1413.
Yamagata, Y., 1966, J.Theor.Biol., 11:495.
Yamagata, T., Hamabe, M., and Iye, M., 1981, Tokyo Obs. Annals., 18:164.
Zel'dovich, B.Ya., Saakyan, D.B., and Sobel'man, I.I., 1977, JETP Lett., 25:94.

NONLOCALITY AND SYMMETRY IN QUANTUM MECHANICS VERSUS LOCALIZABILITY AND SYMMETRY-BREAKING IN PROTOBIOLOGY

Koichiro Matsuno

Nagaoka University of Technology

Nagaoka, 940-21 Japan

1. Introduction

Nonlocal characteristic of quantum mechanics as demonstrated in a simultaneous correlation of polarized particles at different spatial locations[1] suggests that simultaneous specifiability of nonlocal boundary conditions serves as a good approximation to reality[2]. The quantum mechanical equation of motion of the wavefunction preserves the symmetry property observed within the nonlocal boundary conditions. However, nothing propagates at superluminal velocities as Bell's inequalities would imply[3]. One thus observes that if boundary conditions are claimed to be completely and globally specifiable in a simultaneous manner, something would have to be propagated at an infinite velocity in order to guarantee the claimed global specifiability that is of course nonlocal. On the other hand, if simultaneous specifiability of nonlocal boundary conditions is abandoned as it should be and if the locality implying that nothing propagates at superluminal velocities is correctly observed, boundary conditions would have to become necessarily vague and indefinite[4].

We all know that quantum mechanics supplemented by completely specifiable boundary conditions has been truly successful in coping with an extremely wide variety of material processes. Nevertheless, this success alone does not dismiss the presence of such a problem that what would happen if a theoretical artifact of imposing simultaneous specifiability of nonlocal boundary conditions is no more available. At issue is how legitimate it would be to seek nonlocality and the associated symmetry property within boundary conditions.

2. External Measurement and Nonlocality

If boundary conditions are completely controlled with unlimited precision, their nonlocality can be found within the globally simultaneous specifiability or controllability. Moreover, the measurement apparatus employed for checking such a nonlocality also constitutes a part of the boundary conditions to the object to be measured.

Symmetries in Science IV
Edited by B. Gruber and J. H. Yopp
Plenum Press, New York, 1990

In order to see the contribution of measurement apparatus to boundary conditions, let us consider a controlled experiment on measuring three spin coordinates S_x, S_y and S_z of a polarized particle. Here, the eigen-wavefunction of upward oriented spin along the z-direction is ϕ^+_z satisfying

$$S_z \phi^+_z = \frac{\hbar}{2} \phi^+_z$$

and similarly

$$S_z \phi^-_z = -\frac{\hbar}{2} \phi^-_z$$

for downward oriented spin, where $\hbar$ is Planck's constant divided by 2π. Accordingly, linear superposition of the wavefunction ϕ^+_z and ϕ^-_z will give the eigen-wavefunctions of spin polarized along the x- and y-direction such as

$$\begin{cases} S_x (\phi^+_z + \phi^-_z) = \frac{\hbar}{2} (\phi^+_z + \phi^-_z) \\ S_x (\phi^+_z - \phi^-_z) = -\frac{\hbar}{2} (\phi^+_z - \phi^-_z) \end{cases}$$

$$\begin{cases} S_y (\phi^+_z + i\phi^-_z) = \frac{\hbar}{2} (\phi^+_z + i\phi^-_z) \\ S_y (i\phi^+_z + \phi^-_z) = -\frac{\hbar}{2} (i\phi^+_z + \phi^-_z) \end{cases}$$

If a beam of polarized particles is externally measured with regard to the z-component spin and if the meter-reading gives a result of being polarized 100% upward, the spin-wavefunction that has been identified will consist only of the eigen-wavefunction ϕ^+_z, and none of ϕ^-_z. Furthermore, if we imagine such a situation that measurement of the z-component spin is replaced by that of the x-component spin as maintaining all of the other conditions to be the same as previously, the present interchange of measurement apparatus will affect the controllability of experiment only minimally, if at all. This interchange of measurement apparatus will let the eigen-wavefunctions $(\phi^+_z + \phi^-_z)$ and $(\phi^+_z - \phi^-_z)$ for the x-component spin be measurable, instead of ϕ^+_z and ϕ^-_z for the z-component. Accordingly, identity

$$\phi^+_z = \frac{1}{2} (\phi^+_z + \phi^-_z) + \frac{1}{2} (\phi^+_z - \phi^-_z)$$

tells us that measurement of the x-component spin will give 50% being polarized upward and 50% downward along the x-direction.

Let us further suppose that the x-component measurement is followed by the z-component measurement. The spin-wavefunction to be fed into the apparatus measuring the z-component is in the form of either $(\phi^+_z + \phi^-_z)$ or $(\phi^+_z - \phi^-_z)$ because of the involvement of the preceding measurement of the x-component. The subsequent measurement of the z-component will give the meter-reading of 50% being polarized upward and 50% downward along the z-direction.

We thus come up with a superficial paradox or discrepancy such that a beam of polarized particles that have been measured to be 100% polarized upward along the z-direction can be found only 50% polarized upward along the same z-direction if measurement of the x-component spin intervenes and precedes. Measurement of the z-component spin will give different results depending upon whether or not another measurement of the x-component spin intervenes.

However, the difference is more than being superficial. Measurement of the z-component spin implicitly sets such a boundary condition that the spin-wavefunction is a linear superposition of those that are eigen-wavefunctions of the spin operator S_z, instead of those that are the eigen-wavefunctions of S_x or S_y. Likewise, measurement of the x-component spin of the same polarized beam lets the eigen-wavefunction of the z-component spin operator be a linear superposition of the eigen-wavefunctions of the x-component spin operator S_x.

This illustrates that the presence of successive measurements affects what each measurement identifies. Measurements determine a part of the boundary condition under which the quantum-mechanical process to be measured proceeds, especially the nature of the wavefunction in relation to how it is decomposed linearly. Measurement apparatus of any sort specifies the manner of how the wavefunction is linearly decomposed.

The capacity of measurement apparatus for specifying the manner of decomposing the wavefunction will become even more evident if several measurements take place simultaneously at different spatial locations.

Let us suppose a hypothetical experiment to measure the polarization vectors of two particles 1 and 2 in three different ways: two independent measurements of polarization of each particle and simultaneous measurement of two polarizations at the same time. When the polarization of each particle is limited to either plus or minus, the apparatus of measuring particle 1 assumes that the wavefunction $|1\rangle$ is decomposed into two different eigen-wavefunctions as

$$|1\rangle = u^+ |+\rangle + u^- |-\rangle .$$

The apparatus measuring particle 2 also assumes the decomposability

$$|2\rangle = v^+ |+\rangle + v^- |-\rangle .$$

On the other hand, the apparatus that makes simultaneous measurement of two particles possible lets the two-particle wavefunction decomposed into four different eigen-wavefunctions as

$$|1,2\rangle = \alpha |+,+\rangle + \beta |+,-\rangle + \gamma |-,+\rangle + \delta |-,-\rangle .$$

If one chooses photons as polarized particles, it is possible to arrange the equipment so as to let

$$|u^+| = |u^-| = |v^+| = |v^-| = \frac{1}{\sqrt{2}}$$

and

$$\beta = \gamma = 0$$

be observed[1,5]. Then, it follows that although independent measurement of each particle gives 50% chances of being in plus-polarization and 50% chances of being in minus-polarization, there is a 100% parallel correlation of polarization between two photons if both are measured simultaneously. If photon 1 is measured to have plus-polarization, then photon 2 is simultaneously measured to have plus-polarization, and vice versa. This is in fact a simplified demonstration of EPR nonlocality.

The source of nonlocality is however not within something that would propagate at superluminal velocities. Quite to the contrary, the nonlocality is reduced to the manner of decomposing the wavefunction at the apparatus measuring two polarizations simultaneously as in the form

$$|1,2\rangle = \alpha\,|+,+\rangle + \delta\,|-,-\rangle\ .$$

The two-photon wavefunction, when linearly decomposed, does not allow individual two-photon wavefunctions other than those having a 100% parallel correlation of polarization between the two. Such a 100% correlation is an attribute of the measurement apparatus, and by no means the attribute of the measured object.

Any measurement apparatus imposes a specific boundary condition of its own on the object to be measured, especially in the manner of decomposing the impinging wavefunction into the linear superposition of the eigen-wavefunctions for the apparatus. Nonlocality exhibited in a simultaneous correlation extending over different spatial locations just shows a nonlocal property of the eigen-wavefunctions of the apparatus.

Nonlocality of simultaneous spatial correlation of a quantum phenomenon thus reduces to nonlocality of the measurement apparatus. And, nonlocality of the measurement apparatus further reduces to nonlocality of boundary conditions in general, since the apparatus constitutes a part of the boundary condition that makes the physical process to be measured take place. This reduction has successfully been demonstrated experimentally. However, the present experimental demonstration alone does not clarify how the nonlocality of measurement apparatus could be justified, if possible at all.

The nonlocality of measurement apparatus does require that the eigen-wavefunctions of the apparatus are completely specifiable at every moment even though the apparatus is allowed to interact with the object to be measured. This is a necessary consequence from the globally simultaneous specifiability of boundary conditions or their nonlocality. In fact, so long as the complete specifiability of the eigen-wavefunctions of the measurement apparatus is disturbed only infinitesimally during the interaction with the object to be measured, the measurement apparatus can remain external to the measured object as maintaining its nonlocality.

External measurement that makes the eigen-wavefunctions of the measurement apparatus completely specifiable takes its nonlocality for granted. If external measurement were a reality, then the nonlocality of measurement apparatus giving rise to the nonlocality of simultaneous spatial correlation of a quantum phenomenon would also be an undeniable physical reality. Crucial to the matter of nonlocality is whether external measurement can become more than an approximation to reality.

3. Internal Measurement and Localizability

Measurement is ubiquitous among any interacting bodies. In particular, measurement refers to a material process taking place between an arbitrary pair of interacting bodies, in which it is rather customary to call any one of the two the measurement apparatus and the other the measured object[6]. One common denominator of measurement is the capacity of generating mixed quantum states[7] in the sense that measurement irrevocably decomposes the measured object into the mixture of eigen-wavefunctions of the measurement apparatus, though the latter of which can in principle vary its own eigen-wavefunctions through the measurement process itself. Interaction between the measurement apparatus and the measured object renders any one of the two plastic enough to be influenced by the other. The manner of influencing is communicated through successive material interactions. There is no material agent to fully control and specify the manner of communication.

External measurement making the measurement apparatus external to the measured object is thus no more than a theoretical artifact. This artifact forcibly prohibits the measured object from influencing the way of measuring at the measurement apparatus especially in the manner of specifying the eigen-wavefunctins for the aparatus. Unless a prohibitive means is appplied externally, the measurement apparatus remains internal to the measured object in the sense that even the manner of specifying the eigen-wavefunctions for the apparatus depends upon the interaction with the measured object.

Consequently, internal measurement that makes an arbitrary pair of interacting bodies mutually dependent is ubiquitous in physical processes. Internal measurement would reduce to external measurement only at the hypothetical limit of letting the measurement apparatus remain completely specifiable while being influenced by the measured object.

Once internal measurement receives due attention it deserves, the problem of nonlocality will come to gain a new outlook because external measurement equipped with nonlocality of the measurement apparatus is refuted there. Internal measurement defies the globally simultaneous specifiability of boundary conditions and their nonlocality, otherwise it would reduce to external measurement. Even if they are admitted, boundary conditions remain indefinite in their implication and are only partially specifiable at best. Nonlocality of simultaneous spatial correlation of a quantum phenomenon imputed to nonlocality of the measurement apparatuses is foreign to internal measurement.

Internal measurement is unique in maintaining the capacity of only locally successive specifiability, in contrast to globally simultaneous specifiability of external measurement. Internal measurement makes every interaction process local. In fact, locality imputed to Bell's inequality upon sequential propagation of local interactions[3] should be interpreted within the framework of internal measurement, whereas nonlocality imputed to external measurement has nothing to compare with Bell's locality. Any seemingly nonlocal property of a quantum phenomenon is localized within the scheme of internal measurement.

One apparent nonlocality common to any quantum phenomenon is the conservation of energy. If globally simultaneous specifiability were available, the nonlocality of boundary conditions would let the conservation of energy be an attribute of the imposed nonlocality. However, internal measurement defies the involvement of such an external agent to impose a superficial nonlocality.

The absence of any outside agent to control and to specify every detail of the global interaction leads to lack of a material means to completely coordinate the global configuration of interaction in a unique manner. Still, the conservation of energy is observed to be empirically irrefutable. The actualized global configuration of interaction always satisfies the conservation of energy, in spite of the fact that there is no agent to simultaneously coordinate the whole configuration. Instead, internal measurement comes up as a material agent to coordinate the global configuration of interaction, though locally at a time. Uniqueness of the global configuration, however, is lacking when the coordination of interaction is run by internal measurement.

No interaction change initiated at a local region can simultaneously be communicated to the whole region. When the aftereffect of the interaction change initiated at one local region reaches its neighborhood, the interaction configuration in the latter must be so coordinated as to recover energy flow continuity as a local equivalent to the conservation of energy[6]. Successive spill over of local interaction changes for energy flow continuity thus accompanies internal measurement. Beneath this internal measurement lies the local process of materializing conservedness[4] yielded by and entailed by the conservation of energy. Even if there is no external agent claiming the globally simultaneous specifiability or controllability of boundary conditions, internal measurement actualizes the global conservation of energy through the local process of materializing conservedness.

One remarkable property of the process materializing conservedness as a form of internal measurement is its intrinsic irreversibility, as will be seen below. The global configuration of interaction maintained by integrating internal measurements of local character lacks uniqueness in relation to their constituent local configurations because of the absence of the material means to simultaneously coordinate the whole configution. Still, the conservation of energy is and has to be observed a posteriori. Actualization of something that lacks uniqueness of its occurrence in relation to all of the others points to a case of choosing one from many possibilities. The capacity of making choices is thus found to be latent in internal measurement during the transition from the possible to the actual. Intrinsic irreversibility is within the materialistic capacity of making choices enabling the actualization of something that lacks uniqueness of its occurrence, since the choice once made remains irrevocable.

Irreversibility associated with external measurement, on the other hand, is hard to visualize[8], since the quantum mechanical equation of motion, when supplemented by the globally simultaneous specifiability of boundary conditions, yields only a reversible dynamics. The reversibility of quantum mechanics can be saved at the expense of its locality, in which boundary conditions are claimed to be completely specifiable. Nonlocality of boundary conditions makes quantum mechanics reversible. However, once it is recognized that there is no material agent claiming globally simultaneous specifiability, quantum mechanics can make itself free from the overly theoretical commitment to reversible dynamics. Locality of boundary conditions necessarily makes quantum mechanics irreversible.

Nonlocality of simultaneous spatial correlation of a quantum phenomenon reduces to nonlocality of the eigen-wavefunctions for the measurement apparatus. Experimental demonstration of nonlocality of the measurement apparatus is irrefutable. However, the demonstrated nonlocality is not testimony to that there would be an agent claiming globally simultaneous specifiability and unlimited controllability over boundary conditions including even the measurement apparatus.It is of course possible and legitimate in an approximate sense to contrive such an experimental setup that the nonlocality of the measurement apparatus may be preserved to a certain extent against the measured object. There is no question about the significance of what nonlocality experiments suggest. What does matter instead is that the demonstrated nonlocality of the measurement apparatus in experiment does not necessitate the involvement of a material agent claiming globally simultaneous specifiability.

Nonlocality of the measurement apparatus has been proved to be a good approximation to reality in many physical experiments. Locality imputed to Bell's inequality is violated within the scheme of the artificial nonlocality of the measurement apparatus. But, the present approximate nonlocality does not undermine the genuine locality asking no material agent controlling boundary conditions in a completely specifiable manner. The actual locality of quantum mechanics becomes visible when one does away with the nonlocality of measurement apparatus. In fact, one characteristic unique to the locality of quantum mechanics is its intrinsic irreversibility.

4. Symmetry-Breaking in Time and Irreversibility

4.1 Irreversibility in Perspective and Time's Arrow

Before we enter a detailed discussion on the issue of irreversibility, it is needed to make clear our attitude toward a popular question formulated as in the form: Is irreversibility an illusion or a reality? This has been in fact a long standing question, which the inquiry about the nature of time faces. We have already known that there are many arguments for and against time's arrow. In biology and psychology, it seems to us that the relative number of supporters of time's arrow is not marginal compared with that of critics against it. However, in physics, the situation is quite to the contrary. Those critics against time's arrow are overwhelming in physics [9]. Nevertheless, physics undoubtedly serves as a materialistic basis of both biology and psychology. If we fail in finding a materialistic underpinning of time's arrow in physics, both biology and psychology would face a formidable problem of how to justify their time's arrow in phenomenon on a material ground.

It has already been known that there are many different kinds of time's arrow including, for example, thermodynamic, wave-mechanic and K-mesonic ones. The problem is about which one, if any, could deserve time's arrow which remains legitimate and intrinsic in physics.

In discussing physical irreversibility, it has been customary to distinguish between the law of motion and its initial and boundary condition, and to ask which one is responsible for raising the irreversibility. A popular answer to this question is that the source of irreversibility is within boundary conditions while the law of motion is symmetric with time and there is no such thing as time's arrow[9].

In order to see this, let us consider a well-known example of a perfume bottle uncorked. Perfume molecules come out of the bottle and fill the room in time. This motion has been thought to be symmetric in time and the reversed motion of molecules into the bottle is also admissible in principle. But, we have not experienced such an incident even only once. One reason behind is the impossibility of preparing such a boundary condition that all of the molecules in the room enter into a tiny bottle in the course of time development. Boundary conditions serve as an impetus to the superficial irreversibility.

Another example is that we can observe concentric waves propagate outward if we throw a tiny pebble into a pond. But, we have not experienced such an incident that concentric waves excited round the outer periphery of a pond begin to contract and to disappear at its center. One might be able to generate contracting concentric waves by dropping a giant circular hoop into a pond. The problem, however, is its extreme unlikelihood of carrying and dropping such a hoop in the first place. Wave mechanics is symmetric in time like Newton's celestial mechanics and indifferent to whether concentric waves are expanding or contracting. The superficial irreversibility allowing only expanding concentric waves is in the impossibility or the unlikelihood of preparing those boundary conditions that could excite contracting concentric waves.

One more example of this sort is a kettle upon an ice cube. Both our daily experience and thermodynamics tell us that the ice would melt while the water in the kettle becomes colder, in spite of the fact that mechanics alone cannot exclude such an outrageous possibility that the kettle becomes warmer, while the ice gets colder. The superficial irreversibility associated with the kettle becoming colder and the ice warmer is in the extreme rareness of preparing such a boundary condition to the contrary.

All of these three examples suggest such a popular resolution that the source of irreversibility is within boundary conditions, while the law of motion is symmetric with time. A rationale behind this resolution is that the law of motion that is symmetric with time is universal and there is no arrow of time. Boundary conditions, on the other hand, are contingent compared with what is universal, and accordingly, the irreversibility or the arrow of time upon being contingent is thus an illusion. This has been what many physicists and their sympathizers say.

However, we have a different source of irreversibility, which is not found in the former three. In developmental biology, we know for sure that a baby grows into an adult, but, there is no reversed process such that an adult would grow into a baby. If development is a process of making choices as developmental biologists sometimes maintain[10], then the process will exhibit what Watanabe calls the prediction/retrodiction asymmetry[11] because the choice yet to be made is indefinite in its real content, while the one that has already been made is definite in the record.The process of making choices is irreversible because of the irrevocable nature of those choices that have been made, and thus biology witnesses a time's arrow.

Likewise, the similar irreversible process we can find in psychology. We can learn what we have not yet learned, but it is difficult or impossible to unlearn what we have once learned. The arrow of time associated with learning process is in lessening the extent of indefiniteness on the part of the learning subject.

What we have discussed up to this point has already been clearly elucidated by Denbigh[12]. He has made three classes of time. We would like to further reduce them to two classes for the sake of the following discussion. In the superficial irreversibility of the first three examples, the law of motion and its boundary conditions are separable and the law itself is symmetric with time. The irreversibility is contingent upon the boundary conditions available. In contrast, the last two examples admit that the law of motion and its boundary condition are not separable because the boundary condition upon making choices is variable with the progression of making choices. Choices yet to be made are always under the influence of what has been chosen, and learning is also conditioned by what has been learned. In fact, learning is a very special case of making choices because learning is in the process of eliminating what is irrelevant. The boundary condition upon making choices is irreversibly variable in time.

At this point, we wish to raise the following question to ourselves: Namely, does physics allow in itself a room for the law of motion which is inseparable from its boundary condition? The dichotomy we have in mind is whether the law of motion in physics allow only the time-symmetric one or it can also admit the intrinsically irreversible process of making choices on a material ground. If the boundary condition is inseparable from the law of motion, the traditional view letting the boundary condition be contingent would also let the law of motion be contingent, instead of being universal. At issue is whether or not both the law of motion and its boundary condition be contingent.

4.2 Intrinsic Irreversibility

From now on, we shall try to defend the intrinsically irreversible process of making choices on physical ground. The superficial irreversibility originated in boundary conditions alone is fine in its own right. But, what we are going to make is that even in atoms and molecules, the capacity of making choices is ubiquitous and that if we admit that one attribute of mind is the capacity of making choices, then a rudimentary form of mind would also be latent in those atoms and molecules. A clue for this is in the examination of whether or not the law of motion and its boundary condition be separable.

An illustrative example for the present purpose is the constellation of stars in the sky and celestial mechanics. Celestial mechanics following Newton's equations of motion gives a simultaneous expression of all of the heavenly bodies involved, and every star is supposed to detect all of the other stars through its gravitational forces also in a simultaneous manner.

Celestial mechanics thus admits both simultaneous expression of all of the stars and simultaneous detection among themselves as expressed. In essence, it is tantamount to saying that the law of motion is symmetric with time and the boundary condition is separable from the law of motion. However, when we look at the constellation of real stars in the sky, a bit different picture would come up.

Of course, the constellation map of stars depicted as such is a simultaneous expression of all of the stars involved like the equations of motion in celestial mechanics. But we know for sure that the simultaneous expression of stars in the form of constellation map does not imply the simultaneous detection of stars by themselves as depicted in the map. Even if any two stars, which are separated over the distance of, say, one million light years, are simultaneously shown in the map, it takes one million years for either one of the two to detect how the other party is moving.

The idea of the initial condition of all of the stars in the constellation at a given moment is extremely useful in theory as in celestial mechanics. One cannot overemphasize its significance. Nevertheless, one problem remains. In the empirical world, there is no such an agent that can do the job of simultaneous detection of all of the stars as embodied in the initial condition. The idea of the initial condition applies only to a resident in another world of fantasy, who would claim a simultaneous bird's eye view of, say, a universe. Of course, it is not our intention to undervalue the practical utility of the idea of initial conditions as evidenced in physics in general and in engineering and technology in particular. Our point is simply that practical utility cannot and does not undermine reality. What is real is an impossibility of providing the physical basis upon which the idea of initial conditions could be established. Although the initial condition does assume a certain agent that could measure it as such [13], the real difficulty is within the fact that there is no such an agent in the empirical world of physics. The dichotomy of the law of the motion and its initial condition does not count as alleged. An alternative to rely upon should be an incontrovertible empirical fact. A candidate for this role is physical conservation laws because in the latter what is prerequisite is not the idea of initial conditions, but the process of measurement. The separation of the initial condition from the law of motion is fine in the world of fantasy. But, the separation is simply not available in the empirical world of physics.

The difference between celestial mechanics and the constellation map of real stars will become obvious when we consider the conservation of energy in the system. Celestial mechanics takes the energy to be conserved as an attribute of the boundary condition that is separable from the time-symmetric law of motion, because all of the heavenly bodies involved are supposed to detect each other in a simultaneous manner and because the energy is one member to be detected there. On the other hand, the conservation of energy in the constellation of real stars which are, say, more than one million light years apart with each other, does assume an internal process aimed for the very conservation in so far as the empirical principle of conservation is maintained. For it takes time for every star to detect how other stars move in accordance with the conservation of energy and to initiate a change of its interaction with others accordingly.

At this point, one notes that any process of changes does assume something invariant as a reference, otherwise changes themselves could not be referred to. In the present case, the invariant reference is the empirical principle of conservation, and not articulated boundary conditions. One cannot conceive of an invariant and reaction-free boundary condition that can be detected as such by all of the stars involved in the case of real constellation. Although we are quite confident of how the conservation of energy as an attribute of the boundary condition is implemented in celestial mechanics, no matter how difficult it may be technically, we know almost next to nothing about how the conservation of energy is implemented in the constellation of real stars. A key to this observation is within the fact that physical conservation laws are a more fundamental reference than boundary conditions are.

As a matter of fact, when a star is exerted upon by the propagating force field originated elsewhere, we don't know for sure how the force is divided into two parts, one for the acceleration of the star as a whole and the other for the deformation of the shape of the star resulting in a change in the interaction with other stars. This indefiniteness about the division of the role is originated within the lack of invariant reaction-free boundary condition to rely upon for the purpose.

The boundary condition of one star is constituted by all of the other stars and does not remain reaction-free. The detected boundary condition will be reacted upon sooner or later because concurrent changes at two different stars cannot be detected by a third star in a simultaneous manner without any time delay. Reaction-receptive boundary condition is not that kind of boundary condition that remains separable from the law of motion. In fact, only those reaction-receptive boundary conditions are admissible in the empirical world in spite of the fact that an articulation of reaction-free boundary conditions has been proved to be extremely beneficial at least in theory. In other words, the law of motion and its boundary condition are not separable in reality.

Reaction-free boundary condition serves as an unmoved mover,while reaction-receptive boundary condition is no more than a moved mover. What lies beneath this moved mover is physical conservation laws. The reaction-receptive boundary condition of one star in the constellation is in the process such that an aftereffect of detecting variations in the boundary condition is an initiation of new interaction changes with other stars so as to fulfill the conservation of energy a posteriori. Those interaction changes that would violate the principle of the conservation of energy a posteriori are internally eliminated.

The shape of each star is plastic enough to be influenced by the shapes of all of the other stars. However, interaction among real stars takes time, and there is no reaction-free boundary condition to rely upon that can determine the shape of each star in relation to all of the other stars at every moment. Reaction-free boundary condition is too good to be true in the empirical world. Of course, this does not mean that the shapes of stars are arbitrary. The conservation of energy as an irrefutable rule of the game in the empirical world must be observed wherever in the constellation and whenever.

A vehicle of the conservation of energy is interaction. And the effect of the process of interaction is self-constraining in the sense that whatever shape each star may take in relation to all of the other stars with time, those shapes that would violate the conservation of energy a posteriori are eliminated from their realization. Needless to say, the shape of each star that has been actualized remains definite in the sense of being irrevocable in the record. Furthermore, one cannot rely upon a reaction-free boundary condition to identify the realized shape. We are thus led to admit that the process of interaction has the materialistic capacity of picking up a real shape out of the possible by eliminating the impossible.

4.3 Materialistic Capacity of Making Choices

The dichotomy between the law of motion and its initial condition is basically an expression of causality driven by causa efficiens, or simply mechanistic causality. Nevertheless, the process of interaction is more fundamental than the idea of initial condition in that it is interaction, not initial condition, that is responsible for actualizing the empirical conservation laws. This dominance of interaction over initial condition urges us to scrutinize the role of mechanistic causality under a new light. The idea of initial condition deprives the process of interaction of the endogenous capacity for fulfilling the conservation laws internally.

Mechanistic causality and the process of interaction simply do not coexist. One wayout from the present incompatibility, which was originally due to Wheeler and Feynman[14], is to let the interaction be carried by an action-at-a-distance. The interaction mediated by the force of an action-at-a-distance like the static Coulomb force could admittedly be consistent with the operation of mechanistic causality, since the interaction for conservation laws could be adjusted instantaneously and simultaneously in the whole medium and since the initial condition to drive mechanistic causality is identifiable at every instant conservation laws are observed.

However, an action-at-a-distance letting the propagation velocity of interaction change effectively diverge could survive only as an approximation to reality. What is real instead is the interaction mediated by the force of action-through-medium that is propagated at a finite velocity, of course, not exceeding light velocity. Being different from mechanistic causality mediated by the force of action-at-a-distance, the process of interaction is final in the sense that the conservation laws are internally aimed at as being mediated by the force of action-through-medium. What is consistent with the process of interaction is final causality, instead of mechanistic causality. A final cause or causa finalis to drive final causality is physical conservation laws[4].

Final causality latent in the process of interaction is internally selective and regulative in constraining the extent of interaction configuration in the medium so as to fulfill the conservation laws a posteriori. It also lets the local interaction changes for the conservation laws spill over into their neighborhood as causing the similar interaction changes in the latter. Endogenous interaction changes in the mode of final causality are degenerate with those changes to appear later because the indefinite spill over of interaction changes for the conservation laws is internally generative. Henceforth, the process of interaction for the conservation laws is selective only in the sense that it keeps dissolving the internal degeneracy latent in interaction itself with time. The materialistic agent of making choices is degenerate with what it will choose in the future though it is impossible in principle to identify what will be chosen in advance. The dissolution of degeneracy proceeds with the elapse of time.

The materialistic capacity of making choices underlies the process of interaction. Although the term making choices inevitably has an anthropomorphic connotation, we have restricted and shall restrict ourselves to such a usage that it refers only to the materialistic capacity of transforming an a priori indefiniteness into an a posteriori definiteness. If there were reaction-free boundary conditions, physicists would be right when they say that in physics, there is no room for the materialistic capacity of making choices. But, a simple fact is that there is no such thing as reaction-free boundary conditions in reality. What is available instead is reaction-receptive boundary conditions. That is tantamount to saying that the law of motion and its boundary conditions are inseparable.

It is the materialistic capacity of making choices that now comes on the scene when boundary condition recede behind. The materialistic capacity of making choices is, however, not restricted to the constellation of stars. This capacity is ubiquitous in the material world because all of the material units ranging from quarks, nucleons, atoms, molecules and even up to the higher take time when they interact within and with each other.

The point will be made clearer below. Suppose there are ten variables and ten equations of motion to determine their time development. If the initial condition of the ten variables is available, no problem would exist. So far, so good. One the other hand, if the initial condition of one variable is missing for whatever reason, one cannot solve the problem. The way to cope with the present impasse is at least twofold. One is the idealistic dismissal of the problem altogether by maintaining that the problem itself is ill-formulated. The other is a bit mild in the spirit of practical compromise under the guise of either seeking the missing or discarding the variable whose initial condition is missing in order to meet a harsh calibre the idealist sets.

In contrast, what we are trying is the third approach, in which we admit that it is too much to ask that the initial condition of all of the variables could be available. This is because in the empirical world, there is no such an agent that can detect the initial condition as such in an instantaneous and simultaneous manner. Even if the equations of motion are legitimate in their own right, they would be simply dumb unless a necessary and sufficient boundary condition is provided. What we wish to make here is that even if the initial condition of one variable out of ten is missing, those ten variables altogether take care of themselves in order to meet the incontrovertible empirical requirement of the conservation of energy. Internal freedom of making choices is latent there.

The materialistic process of making choices is intrinsically irreversible because of the asymmetry between an a priori indefiniteness and an a posteriori definiteness, thus pointing to the arrow of time in the realm of physics. On the other hand, the law of motion that is separable from its boundary conditions does exhibit the symmetry of time, thus dispensing with time's arrow because of its very definition. However, the presence or the absence of time's arrow is not a matter of definition, but a real question that must be answered. An invariant reference upon which we have tried to answer this question is physical conservation laws that remain empirically legitimate irrespective of the nature of boundary conditions, even if such things would ever exist.

We shall now turn to the aspect of why physicists have long been indifferent to the materialistic capacity of making choices, while physical conservation laws that underlie this capacity have attracted a great attention from physicists.

No physicists argue against the fundamental importance of conservation laws. However, the way physicists have appreciated the importance depends upon the problems they pick up. A most common way of practicing conservation laws in physics is in the mode of ceteris paribus. Even if there are so many molecules as of the order of Avogadro's in a gas, physicists have a pretty good reason to concentrate on the dynamics of, say, two molecules while assuming all of the others are moving in accordance with physical conservation laws. This is in the spirit of perturbation theory that is quite ubiquitous in physical theories. One more prevailing tool is the so-called adiabatic approximation which asks the microscopic movement for conservation laws to be accomplished instantaneously in the whole medium irrespective of the nature of the macroscopic motion to which the microscopic constituent elements are subject.

We all know that practicing conservation laws in the mode of ceteris paribus has been extremely powerful and useful in physics. There is no argument about it. Nevertheless, the physics in the mode of ceteris paribus is not immune to the charge that it has accepted reaction-free boundary conditions in one form or another. What we have recognized is that reaction-free boundary conditions could survive only as an approximation to reality. What is real instead is reaction-receptive boundary conditions. In fact, biologists and psychologists appreciate in their own way the presence of reaction-receptive boundary conditions by pointing out the fact that a subject that has been acted upon by the environment also acts upon the same environment. The materialistic capacity of making choices is a fundamental character of the law of motion that accepts only reaction-receptive boundary conditions.

The materialistic capacity of making choices justifies time's arrow on a material ground because of the materialistic asymmetry between an a priori indefiniteness and an a posteriori definiteness. If one understands the capacity of making choices as an attribute of mind[15], it appears to us that mind underlies time's arrow and that the origin of mind is as old as that of matter. Mind is always with matter. Mind and matter are inseparable, as exhibiting a marked contrast to Cartesian cut claiming to the contrary. Time's arrow in the material world is a phenomenon indicating the faculty of mind latent in matter.

In sum, we have started from the statement that interaction in the material world takes time. What we have then got form the premise is the materialistic capacity of making choices, or time's arrow in the realm of physics. Time's arrow is already latent in the statement that interaction takes time. These two are synonymous and circular. There is no room for contingent reaction-free boundary condition to intervene. Pysicists have followed the tradition of admitting that time's arrow would become an illusion if boundary conditions, though contingent, are a hard reality. This is a conditional statement. What we have made is one more conditional statement, though just the opposite in its implication. Time's arrow would become a hard reality if boundary conditions are an illusion. And we have gone a bit further. Time's arrow is a hard reality because boundary conditions are an illusion.

5. Protoreproduction and Endogenous Symmetry Breaking

5.1 Irreversibility and Evolutionary Process

Symmetry breaking in time is particularly significant in evolutionary process. Evolutionary process persistently undermines articulated boundary conditions in theory while exhibiting an intricate interplay between the law of motion and its boundary conditions.

In fact, empirical observation indicates that evolutionary process constantly renders its products to be the boundary condition for the subsequent production[16]. A material process leading to the emergence of replicating units, in particular, is derived from the process that prior products influence the subsequent production. Initial products come from the production specified by the initial conditions, and the initial products constitute the endogenous boundary condition for the second production, and so on. By following this scheme, products of like kind tend to lead to the production of like kind. As a result, the emergence of the replication of cellular and molecular units turns out to be a consequence of letting the products be the endogenous boundary condition for the subsequent production. Evolution thus comes to render various reaction products to forget about the prior conditions of imposed character and to let the reaction process keep going.

Evolution transforms the boundary condition for the subsequent process in an irreversible manner, at least empirically. The Miller-Urey type abiotic synthesis of amino acids from hydrogen, methane and ammonia, for instance, lets those intermediaries as hydrogen cyanide and formaldehyde serve as the intermediate boundary condition for making amino acis[17]. A mixture of the resulting amino acids now serves as a matrix for producing thermal polymers therefrom[18].

A possible experimental breakthrough for the abiotic synthesis of replicating systems requires a proper scrutiny of how both the law of motion and its boundary conditions develop in a mutually influencing manner. One candidate to draw on for this purpose is the principle of the conservation of energy, matter and the like, because the empirical certitude of the principle is incontrovertible irrespective of whatever the relationship between the law of motion and its boundary conditions may be.

The empirical principle of conservation has at least the three different faces; the state of being conserved, the process of preserving conservedness and process of materializing conservedness[4]. The weakest form among the three is the process materializing conservedness, because the empirical observation is limited only to the measurement that has been done and because the consequence of the process materializing conservedness is simply the observation of conservedness in the record.

In contrast, the state of being conserved implies more than what empirical observation can tell a posteriori. The definability of the state at any moment would inevitably force internal measurement underlying the process communicating conservedness to proceed in the medium at an infinite velocity, in spite of the fact that none of physical processes can be propagated beyond light velocity. The process preserving conservedness simply refers to the temporal transference of the state of being conserved. The temporal transference as a law of motion acting on the state of being conserved lets the latter be the boundary conditions for the process preserving conservedness.

The separability between the law of motion and its boundary conditions is guaranteed in so far as the pair of the process preserving conservedness and the state of being conserved are the case. In fact, the one-to-one temporal mapping of the trajectory in classical mechanics and of the wavefunction in quantum mechanics witnesses this separability. One characteristic consequence of the separability is the preservation of symmetry property, because what has to be conserved is identifiable independently of the law of motion involved. The principle of the conservation of energy, when applied to the state of being conserved, yields a temporally translational symmetry. The symmetry-preserving process is saved, however, at the expense of letting the process communicating conservedness be propagated at an unphysical infinite velocity.

One more theoretical artifact that lets the process communicating conservedness be propagated at an infinite velocity is what one calls the adiabatic approximation asking that when physical process of any type is divided into the two sub-processes, faster and slower, the faster process could be deemed to proceed at an infinite velocity. The adiabatic approximation is ubiquitous in physical theories as pointed out previously. Nevertheless, neither the one-to-one temporal mapping intrinsic to mechanics nor the adiabatic approximation common to most physical theories is competent enough to cope with evolutionary process in which both the law of motion and its boundary conditions are interwoven with each other in a complicate manner.

The process materializing conservedness is intrinsically degenerate with its future development because the impetus to the dynamics is latent within itself. What is more, it constantly dissolves the degeneracy by endogenously eliminating those hypothetical processes that would otherwise fail in accomplishing the principle of conservation a posteriori. The dynamics dissolving the latent degeneracy is fundamentally of one-to-many mapping due to the fact that the process remains indefinite in the forward direction while definite in the backward extrapolation in the sense of being irrevocable. The process materializing conservedness is thus symmetry-breaking and is a form of the law of motion of the one-to-many mapping type.

Evolutionary process, when examined from the physical perspective, requires two ingredients. One is the presence of degeneracy to be dissolved, and the other is the symmetry-breaking process that dissolves the degeneracy. A theoretical prototype of exhibiting these two characters is unstable dynamical systems such as the baker transformation under the premise that a bundle of trajectories is taken as an irreducibly measurable entity[19]. The bundle is infinitely degenerate with trajectories, in which any two of them, infinitesimally separated initially, depart appreciably with time. The bundle dissolves its degeneracy with its development, while the degeneracy still remains inexhaustive. Yet, what has to be examined further is a possible physical justification for letting the bundle of trajectories be an irreducible entity.

One more theoretical prototype exhibiting the two basic characters in evolution is a quantum statistical mechanical system accompanied by frustrations among metastable states further perturbed by disturbances of exogenous origin[20,21]. These frustrations are the source of degeneracy, while exogenous disturbances or errors dissolve the degeneracy. Yet, the statistical mechanical theorizing requires the articulated boundary condition of its own for the origin of both frustrations and exogenous disturbances.

Evolutionary process can be approximated, of course within a limited extent, either by the law of motion of a one-to-one mapping type yielding unstable trajectories or by a specially contrived boundary condition in quantum statistical mechanics. However, approximation cannot undermine reality. Evolution is a particular physical process in which the role of the law of motion of a one-to-many mapping type is correctly appreciated, otherwise evolutionary changes would have to succumb totally to external agents that remain inexplicable.

The fundamental difference between the two types of the law of motion, of one-to-one mapping type and of one-to-many, is in the operation of causality. If the one-to-one temporal mapping is the case, the causality is to operate in the mechanistic mode by letting the separable boundary condition be the mechanicstic cause driving the law of motion. The process preserving conservedness in the future is already explicit at the moment when the boundary condition is identified. In contrast, the law of motion of a one-to-many mapping type operates in the mode of final causality in the sense that the a posteriori principle of conservation serves as the final cause to drive the process materializing conservedness. The tripartite consistency among indefinite degeneracy to be dissolved, endogeneous symmetry-breaking and final causality is latent in the process materializing conservedness.

5.2 Natural Selection

The process materializing conservedness manifests a physical character of protobiological and biological evolution. At the same time, natural selection serves as the biological basis for evolutionary process. We thus come across the problem of how the process materializing conservedness and natural selection conform with each other.

Natural selection refers in its effect[22] to differential reproduction under resource limitation. Resource limitation is an expression of the conservation of matter. The principle of the conservation of matter, however, differs in its implication depending upon whether the underlying causality is mechanistic or final. If the process communicating resource limitation or conservedness would proceed at an infinite velocity, mechanistic causality would come to prevail in such a manner that resource limitation serves as a mechanistic cause to drive natural selection. For the limitation is observed whenever and wherever. Resource limitation fixed as a boundary condition being separable from the law of motion in fact makes what raises differential reproduction to be a mechanistic cause for natural selection, because differential reproduction points to how the limitation set as a materialistic antecedent would actually be implemented.

Resource limitation in the mode of mechanistic causality would accordingly affect, in addition to effecting, differential reproduction. Natural selection affecting[23] differential reproduction could follow. Evolutionary changes in the scheme of mechanistic causality would necessarily have to be originated in external agents that may control boundary conditions in an inexplicable manner. Nevertheless, the mechanistic causality is based upon the physically unattainable premise asking the process communicating resource limitation to proceed at an infinite velocity.

The process materializing conservedness implies the communication of resource limitation proceeding at a finite velocity. Accordingly, the process lets resource limitation operate in the mode of final causality such that each of replicating material units adjusts its interaction with other so as to fulfill the principle of resource limitation a posteriori. Internal process for physical conservation laws is final, while the process imposed upon these conservation laws is mechanistic by letting the laws themselves be separable from the dynamics whatever it may be.

Resource limitation in the mode of final causality is thus equivalent to resource exploitation at each replicating unit. Resource exploitation indicates the latent capacity of material units not only to be influenced by but also to influence their exterior in order to meet the a posteriori principle of resource limitation. The final mode of causality lets natural selection simply effect differential reproduction, being different from the case of mechanistic causality affecting it.

Natural selection effecting differential reproduction operates through dissolving the degeneracy latent within evolutionary precursors. Both the genesis of de novo variations and the self-limitation are associated with the successive dissolution of degeneracy. It would thus be inappropriate to say that natural selection could act upon variations. Natural selection and variations are inseparable in the legitimate scheme of final causality. Although it might be said that natural selection is to act upon variations if the latter are prepared as an attribute of boundary conditions independently of mechanism effecting differential reproduction, such a mechanistic scheme could survive only under a theoretical articulation letting the very mechanism of natural selection, whatever it may be, also be originated in exogenous boundary conditions. Natural selection in the mode of mechanistic causality would let both the mechanism of selection and those to be selected be of exogenous origin in a mutually independent manner, no matter how unlikely their independence may be.

Natural selection in the mode of final causality, on the other hand, refers to no more than the capacity of resource exploitation on the part of participating material units whether molecular, cellular or organismic. Differential reproduction is only an effect of natural selection, and not its cause. In this regard, neo-Darwinian natural selection acting upon phenotype, while variations upon genotype, is mechanistic in its causation because of the separability between the mechanism of selection and the genesis of variation. An advantage of mechanistic causality is in its capacity to assimilate the logical antecedent with the mechanistic causative factor. However, resource exploitation acting in the mode of final causality witnesses just the opposite, namely the logical antecedent acting as a final factor.

The inseparability between natural selection and the process generating variations would necessarily invoke a bilateral causation between genotype and phenotype, a resurrection of Lamarckian ghost or a neo-Lamarckism[24]. The neo-Darwinian scheme has enjoyed its theoretical clearness because of the underlying mechanistic causality and has also been factually supported by the Weismannian separation between soma and germen[24]. In contrast, the Lamarckian reciprocity between genotype and phenotype has long been denigrated and charged as lacking its analytical clarity, to say the least.

The problem to be raised at this point is whether the Weismannian separation could be factually complete to such an extent as the neo-Darwinian premise would ask in theory. The process materializing conservedness that yields natural selection acting in the mode of final causality urges us to re-examine or even to salvage the once denigrated neo-Lamarckian inseparability between natural selection and variations, though the analytical clarity as enjoyed by the neo-Darwinian counterpart may not be feasible.

5.3 Protoreproduction

Natural selection effecting differential reproduction does not necessarily presume replicating units, being contrary to the hypothetical case of affecting it. Resource limitation in the mode of final causality, or resource exploitation in short, can be responsible not only for effecting differential reproduction among replicating units but also for raising those units ab initio. Resource exploitation in progress leaves the resource exploitation with the greater competence because the process operating in the mode of final causality is degenerate with the possible future processes and because it constantly dissolves the degeneracy as leaving the one that can be capable for the successive resource exploitation. The material units that are more competent in resource exploitation can emerge through the very resource exploitation. The emergence of replicating units with the greater competence for resource exploitation is in fact an instance of enhancing resource exploitation by leaving the similar progeny with the similar competence.

Resource limitation in the mode of mechanistic causality alone cannot raise replicating molecules unless it is further supplemented by external agents to raise them. In contrast, resource limitation in the mode of final causality can intrinsically be pregnant with possible replicating molecules because of the capacity of resource exploitation on the part of molecules and their aggregates. Final causality is in fact observed in the materialistic process aimed for physical conservation laws. Unless it is willfully dethroned by contriving such an artifact that experimenters or theoreticians to a lesser extent could control every cause of evolutionary process, final causality firmly grounded upon material basis surely survives.

What is more, material aggregation or self-assembly is also an instance of resource exploitation. Material self-assembly and natural selection effecting differential reproduction thus share the same functional underpinning of resource exploitation. The only difference is its expression. The process materializing conservedness renders the emergence of replicating units and their differential reproduction to be a dissolved form of the degeneracy latent in the material self-assembly at the protobiological level. Material self-assembly at the molecular and the protocellular levels can raise protoreproductive material units[25)], though laboratory experiments have not yet been conclusive for the emergence of replicating molecules through resource exploitation.

Material units participating in resource exploitation take in not only molecules but also energy to activate those molecules to let them be involved in molecular self-assembly. Energy exploitation as a particular mode of resource exploitation is an endogeneous process not solely controlled by external conditions. In fact, energy exploitation that can be more competent succeeds the lesser one in an endogenous manner. The present endogeneity sheds some light upon how molecular replication could get started.

If the amount of incident energy required for molecular replication upon a template as with a possible abiotic RNA replication[26,27)] is controlled by exogenous boundary conditions[28)], it could be at most a rare coincidence that the incident energy happens to be the right amount required internally. Otherwise molecular replication could not survive even if once started. In contrast, if molecular replication was from the start equipped with the mechanism that is capable of energy exploitation[29)], the energy intake could be controlled endogenously so as to meet the level required internally. Molecular replication could survive under such a circumstance.

Energy exploitation is in fact actualized in the process materializing conservedness of energy or, energy equilibration, that refers to the endogenous adjustment of interaction so as to meet the conservation of energy a posteriori. The similar endogeneity holds for material resource exploitation in the form of material flow equilibration that refers to the process actualizing material flow continuity a posteriori. Equilibration, whether in energy or in material flow, is in the endogeneous capacity of material units to act upon their exterior.

A possible mechanism of energy exploitation that could emerge at the protobiological stage is oxidative phosphorylation[30)]. Bilayer spherules made of phospholipids or phosphlipid like materials[31,32)] could serve as material units to implement such an energy exploitation mechanism.

Fundamental to the origin and preservation of molecular replication is the control of both energy and molecular intakes. The intake control has be endogeneous in the sense that it needs to be adjusted in response to what is going on inside. A program of laboratory experiment intended for the internal genesis of endogeneous controllability has actually been proposed[33]. Against this endogenous controllability, there could also be the exogenous counterpart. It is possible within the currently available knowledge to conceive or even to exogenously set up an experimental boundary condition that could simulate the controllability endogenous to material units involved in resource exploitation of their own. In fact, layer replications of montmorillonite up to 20-23rd generation have been observed in an externally controlled experimentation[34]. However, the origin and evolution of the endogeneous controllability is not found within the exogenous simulation. At stake is how the endogeneous controllability of resource exploitation could develop with the exploitation. The underlying rationale is the law of motion in the mode of final causality. Protoreproduction, whether molecular or cellular, manifests how resource exploitation evolved especially at an early evolutionary stage.

Significant to evolutionary process is an appraisal of final causality in physical processes, irrespective of the fact that it has long been demeaned[35] in the face of the overwhelming mechanistic causality. When one views evolutionary process as a temporal transformation of the control mechanism of material self-assembly, the time-honored mechanistic causality alone could not be good enough because it would have to seek the origin of variations in the mechanism exclusively in external agents that would remain inexplicable. In contrast, final causality manifested in the internal process aimed for physical conservation laws lets resource exploitation be variable through resource exploitation. The control mechanism of material self-assembly can vary through the very process of control. The execution program varies endogenously with its execution[36].

Earlier evolutionary mechanism is infinitely degenerate with those to appear later. Evolution is a process dissolving the degeneracy. The evolutionary emergence of replicating molecules is a form of dissolving the degeneracy latent in the control mechanism of non-replicating molecular self-assembly. Final causality intrinsic to the process materializing conservedness in fact provides the material basis of both the degeneracy to be dissolved and its dissolution in the form of endogenous symmetry-greaking.

Protoreproduction happens to be a quantum mechanical phenomenon that requires neither nonlocality of boundary conditions nor symmetry-preservation. In fact, one mode of practicing quantum mechanics without being supplemented by nonlocality and symmetry is to properly appreciate final causality deduced from the process entailed by and leading to physical conservation laws. Localizability rejecting instantaneous communication over nonlocal distances and the associated symmetry-breaking are prerequisite to protobiology as a form of quantum mechanics that can give birth to biology.

5.4 Information Dynamics

Endogenous symmetry-breaking in protobiology and accordingly, in biology also, urges us to re-examine the superficial nonlocality of quantum mechanics requiring a symmetric property for its own sake. One of the example for the study of nonlocality or localizability in quantum mechanics is a DNA molecule because it plays a crucial role in biology while at the same time being a quantum mechanical molecule maintained by electrostatic interaction[37].

A major characteristic of a DNA molecule in evolution is its nucleotide substitution identified as point mutation. The gene coding cytochrome C has been measured to have roughly 5 nucleotide substitutions per 100 codons per 100MY(million years)[38]. Since there are four alternations of A,T,G and C at each nucleotide substitution, the process of point mutation comes to enhance specificity in the sense that it has the capacity of realizing one out of the four alternative.

If there are two alternatives with equal likelihood of occurrence and if a choice is made by whatever means over a unit time interval, the choice based upon the capacity of enhancing specificity comes to generate information at the rate of one bit per unit time. Here, information is understood to be a capacity of enhancing specificity or making choices. Information to be transcribed as in the contemporary transcription from DNA to protein is not our concern here. In fact, information to be transcribed is simply a product of information as a capacity of enhancing specificity. Information as a capacity of enhancing specificity is process-information whereas information to be transcribed is product-information. We shall concern ourselves only with process-information unless mentioned otherwise.

The above consideration, when applied to nucleotide substitution of the gene coding cytochrome C, can readily lead to the rate of information generation to be 10^{-17} bits per second per nucleotide-site. Here, we have employed the empirical observation that each nucleotide substitution out of four alternatives of A,T,G and C would occur with equal likelihood. Furthermore, since nucleotide-site has only one degree of freedom whose value take one out of four of A,T,G and C, the information generation rate at the gene coding cytochrome C is found to be

$$10^{-17} \text{ bits/s/degree-of-freedom.}$$

A similar estimation can be done for those genes coding other proteins such as myoglobin, α hemoglobin, β hemoglobin, α lens crystallin, fibrinopeptide A and B, carbonic anhydrase and copper-zinc superoxide dismutase[38,39]. The associated information generation rate falls within the range $1\sim6\times10^{-17}$ bits per second per degree of freedom.

Nonzero information generation rate has been derived purely on an empirical basis without employing any kind of theoretical artifact asking for nonlocality. However, if one demands a definite boundary condition for a DNA molecule, which is of course nonlocal in its nature, there could be no room for nucleotide substitution because of the lack of any indefiniteness about the imposed boundary condition. The information generation rate would have to vanish. The imposed nonlocality would destroy the capacity of nucleotide substitution in a DNA altogether. But empirical facts simply invalidate such a nonlocality.

We now notice that it is always possible to theoretically approximate such a small information generation rate of order of 10^{-17} bits per second per degree of freedom to be effectively zero. But this does not prove nonlocality of quantum mechanics. Quite to the contrary, nonlocality demanding the information generation rate to vanish is no more than an approximation to quantum mechanics that admits nonvanishing information generation rate and the associated localizability.

Information generation is not limited to the one due to electrostatic interaction as exhibited in the nucleotide substitution in a DNA. Electrophysiological interaction acting in plants can also generate information. In particular the turnover rate of organic carbons in plants is found to be $0.1\sim0.15$ per year[40]. This means that if one takes any two carbon atoms in a plant, the probability that any one of the two is replaced by a new one from the outside within about three years is roughly one half. This gives the information generation rate of 10^{-8} bits per second per carbon. Since each carbon appearing in the turnover process represents only one degree of freedom, the rate leads to

$$10^{-8} \text{ bits/s/degree-of-freedom.}$$

One more interaction that can generate information is mechanochemical one as seen in cell motility[41]. Take, for instance, a dynein-tubulin complex of a flagellum[42]. The relationship between the displacement of the medium and the force acting there is of one-to-many correspondence. There are many possibilities in anticipation, but unique in retrospect. Transformation from many in the possible to one in the actual is the process of enhancing specificity and thus generates information. The result yields

10^2 bits/s/degree-of-freedom.

We have seen three different processes of information generation. The characteristic propagation velocity of changes in interaction is $3x10^8$ m/s for electrostatic, 10 m/s for electrophysiological and 10^{-5} m/s for mechanochemical interaction[43]. If the propagation velocity of interaction changes diverges, every interacting element would come to detect how all of the others move in an instantaneous manner. Nonlocality as a capacity of detecting the global character instantaneously would thus survive. The information generation rate would vanish there. However, nothing propagates at superluminal velocities. The rate remains nonzero even at the propagation velocity approaching the limit of light velocity as exhibited in the case of electrostatic interaction.

Furthermore, the information generation rate tends to increase as the propagation velocity of interaction changes decreases. This is because the extent to which each interacting element is simultaneously constrained is lessened as the velocity at which each communicates with others decreases. This qualitative characteristic of inverse proportionality between the propagation velocity of interaction changes and the information generation rate has been confirmed on an empirical basis.

Above all, what is significant in protobiology and in biology is a coherent interplay between the two processes of information generation, fast and slow. The slow information generation process due to electrostatic interaction assumes the role of causing evolutionary changes. On the other hand, the fast processes due to electrophysiological and mechanochemical interaction represent morphological process in development. The slow information generation process is ubiquitous even in the abiotic realm. Unless it interferes with the fast process of information generation, the slow process can be approximated to generate no information as physicists take it for granted in the nonbiological sector.

A uniqueness of biology is seen in the nonvanishing rate of slow information generation and its incorporation into fast information generation processes. The interplay between fast and slow information generation processes is certainly a case of quantum mechanics demonstrating its localizability.

References

1 Aspect, A., Grangier, P., and Roger, G., 1981. Experimental tests of realistic local thories via Bell's theorem, Phys. Rev. Lett. 47, 460.

2 Matsuno, K., 1989. Nonlocality and localizability in quantum mechanics, Ann.Fond.Louis de Broglie 14, 233.

3 Bell, J. S., 1989. Speakable and Unspeakable in Quantum Mechanics 1987. (Cambridge University Press, London).

4 Matsuno, K., 1989. Protobiology: Physical Basis of Biology (CRC Press, Boca Raton, Florida).

5 Aspect, A., and Grangier, P., 1983. Experiment on Einstein-Podolsky-Rosen type correlations with pairs of visible photons, in Proc. Int. Symp. on Foundation of Quantum Mechanics, Kamefuchi, S., et al Eds. (Phys. Soc. Japan, Tokyo), p. 214.

6 Matsuno, K., 1985. How can quantum mechanics of material evolution be possible? : symmetry and symmetry-breaking in protobiological evolution, BioSystems 17, 179.

7 Conrad, M., 1983. Adaptability: The Significance of Variability from Molecule to Ecosystem (Plenum, New York), chap. 2.

8 Lochak, G., 1981. Irreversibility in physics: reflections on the

evolution of ideas in mechanics and the actual crisis in physics, Found. Phys. 11, 593.
9 Mehlberg, H., 1980. Time, Causality, and the Quantum Theory, Vol.1 (D. Reidel, Boston), chaps. 7 and 9.
10 Slack, J. M. W., 1983. From Egg to Embryo: Determinative Events in Early Development (Cambridge University Press, Cambridge) chaps. 1 and 2.
11 Watanabe, S., 1966. Time and the probabilistic view of the world, in The Voices of Time, Fraser, J. T., Ed. (Braziller, New York) p. 527.
12 Denbigh, K. G., 1981. Three Concepts of Time (Springer-Verlag, Berlin), chaps. 6, 7 and 8.
13 Schulman, L. S., 1986. Deterministic quantum evolution through modification of the hypotheses of statistical mechanics, J. Stat. Phys. 42, 689.
14 Wheeler, J. A., and Feynman, R. P., 1949. Classical electrodynamics in terms of direct interparticle action, Rev. Mod. Phys. 21, 425.
15 Matsuno, K., 1984. Determinism and freedom in early evolution, in Individuality and Determinism: Chemical and Biological Bases, Fox, S. W., Ed. (Plenum, New York), p. 203.
16 Matsuno, K., 1984. Is matter inanimate: protobiological information from within, Origins Life 14, 489.
17 Miller,S., and Orgel, L. E., 1974. The Origin of Life on the Earth (Prentice-Hall, Englewood Cliffs, New Jersey).
18 Fox, S. W., 1988. The Emergence of Life: Darwinian Evolution from the Inside (Basic Books, New York).
19 Courbage, M., and Prigogine, I., 1983. Intrinsic randomness and intrinsic irreversibility in classical dynamical systems, Proc. Natl. Acad. Sci. USA 80, 2412.
20 Anderson, P. W., 1983, Suggested model for prebiotic evolution: The use of chaos, Proc. Natl. Acad. Sci. USA 80, 3386.
21 Stein, L. D., 1984. A model for the origin of biological information, Int. J. Quant. Chem. QBS-11, 73.
22 Vrba, E. S., and Eldredge, N., 1984. Individuals, hierarchies and processes: towards a more complete evolutionary theory, Paleobiology 19, 146.
23 Bernstein, H., Byerly, H. C., F.Hopf, F., Michod, R. A., and Vemulapalli, G. K., 1983. The Darwinian dynamic, Quart. Rev. Biol. 58, 185.
24 Goodwin, B. C., 1984. A relational or field theory of reproduction and its evolutionary implications, in Beyond Neo-Darwinism, Ho, M. W., and Saunders, P. T., Eds. (Academic Press, London), p. 219.
25 Matsuno, K., 1984. Open systems and the origin of protoreproductive units, in Beyond Neo-Darwimism, Ho, M. W., and Saunders, P. T., Eds. (Academic Press, London), p. 61.
26 Lohrmann, R., Bridson, P. K., and Orgel, L. E., 1980. Efficient metal-ion catalyzed template-direct oligonucleotide synthesis, Science 208, 1464.
27 Biebricher, C. K., 1983. Darwinian selection of self-replicating RNA molecules, in Evolutionary Biology, Vol. 16, Hechet, M. K., Wallace, B., and Prance, C. T., Eds. (Plenum, New York), p. 1.
28 Eigen, M., and Schuster, P., 1982. Stages of emergence of life - five principles of early evolution, J. Mol. Evol. 19, 47.
29 Morowitz, H., 1978. Foundations of Bioenergetics (Academic Press, New York).
30 Koch, A. L., 1985. Primeval cell: possible energy-generating and cell-division mechanisms, J. Mol. Evol. 21, 270.
31 J.Douglas, J., 1984. Hypothetical entropy-driven mechanism for self-regulation of the size and division of primitive cells suggesting the origin and nature of mesosomes, J. Theor. Biol. 109, 475.
32 Deamer, D. W., and Oro, J., 1980. Role of lipids in prebiotic

structures, BioSystems 12, 167.
33 Pattee, H. H., 1979. The complementarity principle and the origin of macromolecular information, BioSystems 11, 259.
34 Weiss, A., 1981; Replication and evolution in inorganic systems, Angew. Chem. Int. Ed. Engl. 20, 850.
35 Rosen, R., 1985. The physics of complexity, Syst. Res. 2, 171.
36 Mayr, E., 1980. Prologue: some thoughts on the history of the evolutionary synthesis, in The Evolutionary Synthesis, Mayr, E., and Provine, W. B., Eds. (Harvard University Press, Cambridge), p. 1.
37 Pullman, A., and Pullman, B., 1981. in Chemical Applications of Atomic and Molecular Electrostatic Potentials, Politzer, P., and Truhlar, D. G., Eds. (Plenum, New York), p. 381.
38 Goodman, M., 1981. Decoding the pattern of protein evolution, Progr. Biophys. Molec. Biol. 37, 105.
39 Ayala, F. J., 1986. On the virtues and pitfalls of the molecular evolutionary clock, J. Hered. 77, 226.
40 Odum, E. P., 1971. Fundamentals of Ecology, 3rd Ed. (Saunders, W. B., Philadelphia).
41 Matsuno, K., 1989. Cell motility: an interplay between local and nonlocal measurement, BioSystems 22, 117.
42 Hiramoto, Y., and Baba, S. A., 1978. A quantitative analysis of flagellar movement in echinoderm spermatozoa, J. Exp. Biol. 76, 85.
43 Harada, Y., Noguchi, A., Kishino, A., and Yanagida, T., 1987. Sliding movement of single actin filaments on one-headed myosin filaments, Nature 326, 805.

EVOLUTIONARY MEANINGS OF THE PRIMARY AND SECONDARY STRUCTURES OF THE "UR-RNA", A PRIMITIVE POSSIBLY SELF-REPLICATING RIBO-ORGANISM COMMONLY ANCESTRAL TO tRNAs, 5S-rRNA AND VIRUSOIDS

Koji OHNISHI

Department of Biology, Faculty of Science, Niigata University

Ikarashi-2, Niigata 950-21, Japan

1. INTRODUCTION

Recent advances in the study of catalytic RNAs have allowed us to consider that the first self-replicating organism was almost undoubtedly an RNA without protein-encoding informations (Watson et al., 1987; Orgel, 1987; Schwartz et al., 1987; Cech, 1989a; Joice, 1989). The class I intron of *Tetrahymena* pre-rRNA, virusoids, viroids, RNA portion (M1 RNA) of ribonuclease P and the delta element of hepatitis B virus are representatives of these catalytic RNAs (See reviews in Cech, 1989b and Diener, 1987a).

Viroids and virusoids (See Diener, 1987b,c for review) are known to be small, circular, single-stranded, highly base-paired rod-like RNA pathogens of plants (for viroids, see Haseloff and Symons, 1981; Keese and Symons, 1987; and for virusoids, see Keese et al., 1983; Francki, 1987), and to have a self-cleaving ribozyme activity (for viroids, see Saenger, 1987; Branch et al., 1988; and for virusoids, see Forster and Symons, 1987a, b; Forster et al., 1987). Virusoids are satellite RNAs because they are dependent on viruses for replication and encapsilation, whereas viroids are sole infectious units. Two alternative rolling-circle models, asymmetric (Branch et al., 1981) and symmetric (Branch and Robertson, 1984), have been proposed for viroid replication (Saenger, 1987). Branch et al. (1988) have recently demonstrated that potato spindle tuber viroid appears to use the asymmetric model in which a circular plus-strand is copied to give a long "multimeric linear minus-strand", and the minus-strand is further copied to a multimeric plus-strand precursor, and the plus-strand precursor is finally cleaved and circularized to give a new viroid.

Recent findings of a close homology among tryptophanyl-tRNA ($tRNA^{Trp}$), 5S rRNA and virusoids have confirmed that all of these RNAs and other tRNAs evolved from a primitive RNA with about a length of 88 bases (Ohnishi, 1989a, b, c). This ancestral RNA is called "ur-RNA", and can be considered to be a possibly self-replicating (auto-catalytic) or mutually catalytic living RNA, or in other words, a primitive ribo-organism (hereafter we denote "ribo-organism" by "RO") (Ohnishi, 1989a, b, c). These findings seem to suggest important keys to solve the origin and evolution of genetic apparatus as well as the origin of various types of ribozymes including virusoids/viroids and M1 RNA.

On the other hand, the "primodial gene theory" (Ohnishi, 1986a, b) has

Symmetries in Science IV
Edited by B. Gruber and J. H. Yopp
Plenum Press, New York, 1990

concluded that ca. 47-aa-long domains of co-origin are fundamental repeating units of house-keeping enzymes including adenylate kinase (AK), F_1-ATPase epsilon and gamma subunits, some aminoacyl-tRNA synthetases and an E. coli ribosomal protein L34. How the primodial gene encoding an ancestral ca.47-aa-long primitive protein emerged in relation to peptide-non-encoding ROs is an important basic question remaining to be answered. This problem is also very important from the aspect of the origin of triplet codon-anticodon complementarity encoding specific amino acids.

Here I will review my recent studies and present some new findings on the ur-RNA and related topics on the origin and evolution of early RNAs or ROs and of early genetic apparatus. An important viewpoint to be stressed is the symbiotic theory on the origin of the cell, which hypothesizes that the cell machinery evolved as a co-operative symbiotic society of populations of various ROs including tRNAs, rRNAs, M1 RNA, and mRNAs, and proteins evolved as cultural products or machines (or tools) [or a kind of "meme", if we use Dawkin's (1976, 1982) terminology] in this cooperative society. The evolutionary meanings of these findings will also be discussed from the aspect of the evolution of hierarchical structure of individuality and generalized culture. Origin of triplet anti-codon of tRNA will be considered in the last section.

2. PREDICTION OF PRIMARY AND SECONDARY STRUCTURES OF UR-RNA

2.1. Homology relationship among $tRNA^{Trp}$, 5S rRNA, and virusoids

In order to elucidate evolutionary relationship of contemporary RNAs to early ROs, homology relationship among the primary sequences of virusoids, tRNAs and 5S rRNAs were extensively searched. The primary sequence of the circular satellite RNA (RNA 2, 324 bases) of lucern transient streak virus (LTSV) strain A (LTSV-A) (Keese et al., 1983) was used in this study as a representative of virusoids, since it's molecular weight is the smallest of hitherto sequenced virusoids. The resulting alignment is shown in Fig. 1. Homology levels in base sequence alignments were evaluated by computing

$$P_{nuc}(m,n) = \Sigma_{i=m}^{n} [n!/(i!(n-i)!)](1/4)^i(3/4)^{n-i}, \qquad \text{[Eq. 1]}$$

which denotes the probability that two randomly selected n-base-long polynucleotides share identical bases at m or more aligned base positions under the assumption of equal occurrences of 4 different bases (Ohnishi, 1984a,b, c). Eq. 1 is based on binominal distribution.

Base matches and P_{nuc} values observed in the alignment in Fig. 1 are summarized in Table 1. From these, we find the followings:

(1) The entire molecule (bases 1-76) of E. coli (EC) $tRNA^{Trp}$ is highly homologous to the 5'-terminal two thirds (bases 1-76) of Bacillus stearothermophilus (BST) 5S rRNA, with 43 % base match (P_{nuc} = .0011). The EC tRNAs for Phe, Ile and Met also show considerable homology to 5S rRNA, whereas EC $tRNA^{Ala}$ shows only 24% base match. Thus $tRNA^{Trp}$ sequence seems to be more conservative than any other contemporary tRNA sequence.

(2) The remaining one third (bases 77-117) of the BST 5S rRNA is again homologous to the 5'-terminal half of the EC $tRNA^{Trp}$, with 45% match (P_{nuc} = .0063). 5S rRNA, therefore, first emerged most probably by tandem duplication of ancestral tRNA or its close homologue.

(3) The terminal highly base-paired portion (bases 285-324,1-42) of the LTSV-A RNA 2 virusoid is highly homologous both to EC $tRNA^{Trp}$ (55%, P_{nuc} = .28 X 10^{-6}), to EC 5S rRNA (1-76) and (77-117) with 42% and 51%, P_{nuc} = .0025 and .00037, respectively.

Accordingly, it can be concluded that the close similarities among $tRNA^{Trp}$, 5S rRNA and virusoid aligned in Fig. 1 are genuine homologies as resulted from their common ancestry (Ohnishi 1989a, b, c).

```
                                           [-- B --]         [---C--]
                              +   +        ++  + +++    ++  +     +
5S rRNA (BST)    5' (1)    ---CCUAGUGACAAUAGC-GG-AGAG---GAAACAC--CCGUU-CCCAUC-  (39)
 (bases 1-76)                 :            :: :: :::    :    ::  ::::    :     )
                 3'(76)/AUGGUAGCCG---------CGACC-UCU----CGAAUUGAAGGCAC-AAG-CC-  (40)
                            + + ++          +  + +++    +  +         +  ++ +

                            [Acc.]
                            ++  +++ ++         + +++    +   +++   ++     +
                                  **      *   ** ***    * * *     *       * *
tRNATrp (EC):    5' (1)     AGGGGCGØAGUUCAAØØ-GGØAGA----GCACCGG--UØUCC-AØAACC-  (41)
                            ::::::            :: :::    :   ::    :::: :: :::  )
                 3'(76) ACCGUCCCCG------------CC-UCU----CUGAGCØ--ØGAGGGUUØUGGG  (42)
                           **  ***            ** ***    *  *      *      *

                          + ++  ++             + +++    ++ +++   + +     ++

                              *   *        **  * ***    **  *     *
LTSV-A RNA 2        (1)   -GAGCUGCGCAG--GGGGC-CG-AGAUUUUGAUUCGG--ACUGG--UAGGA-/ (42)
 (virusoid)             (   :::::: : :    : :  : ::::   :: ::::  : :::  :  ::
                  (324)   CUUCGACG-GAC-CUACUG-UC-UCUAU--CU-AGCC--U-ACC-GAGUCU-/(285)
                            * * **          *  * ***    *  *         *  ** *

ur-RNA           5' (1)   -gAGCYGCGYAGynRRRGC-NG-AGANuuuGANYCGG--NCUNN-nYARNN-  (44)
                            :::;:: , :    ,,: ,: :::,   :: ,::,  ;::;; ,; ;;;  )
                 3'(88) ayCNUCGNCG-gac-cuacYG-YC-UCUau--CUrAGCY--UgANC-RAGUCY-  (45)

5S rRNA (BST)    5'(77)  /GUUGGGGC-CAG--CGCCC-CU-GCAA---GAGUAGG--UCGUUGCUAGGC-  3'
 (bases 77-117)              +  ++ +++   +  + +    +    ++ + ++   +     ++++   (117)
                                            *      *    **        **** *    *
```

Fig. 1. Alignment of base sequences of 5S rRNA (BST = *Bacillus stearothermophilus*, major), tRNA^{Trp}(EC = E. coli) and LTSV-A RNA 2. LTSV-A is an isolate of lucern transient streak virus (LTSV), and RNA 2 is a virusoid dependent on the LTSV. G-C and A-U complementary base-pairs are indicated by ":". ")" and "(" indicate continuity of sequence. Base matches to tRNA^{Trp}, LTSV-A RNA 2 and 5S rRNA (basea 1-76) are indicated by "_", "+" and "*", respectively. ";" and "," in ur-RNA are positions where complementary base-pair can be inferred with considerable plausibility. [B], [C] and [Acc.] are the regions of B stem and C stem of 5S rRNA and the acceptor stem of tRNA, respectively. Modified residues in tRNA^{Trp} are over-printed by slash ("/"). Bases in lower case letter are estimated by limited data, and R = A or G, Y = U or C, N = R or Y. Residue 1 of LTSV RNA 2 corresponds to the residue 2 in Keese *et al.*'s original numbering. Sequence data are from: LTSV-A RNA 2 = Keese *et al.*, 1983; 5S rRNA = Marotta *et al.*, 1976; tRNA^{Trp} = Hirsch, 1971.

Table 1. Percent base matches (%) (lower left) and $P_{nuc}(m,n)$ (upper right)* in the alignment of Fig. 1.

(1) 5S RNA (BST) 1-76		.0011	.0025	.56
(2) tRNATrp (EC)	42.6%(29/68)		$.28 \times 10^{-6}$	.0063
(3) LTSV-A 2 RNA 2	41.5 (27/65)	54.5(36/66)		.00037
(4) 5S rRNA(BST) 77-117	27.7 (10/36)	44.7(17/38)	51.3(20/39)	
	(1)	(2)	(3)	(4)

* $P_{nuc}(m,n)$ is the probability by chance for giving m or more base matches in n aligned positions assuming equal occurrence of different 4 bases. m/n are given in parentheses. Gap sites are ignored in calculation of percent matches.

2.2. Prediction of ur-RNA structure

This common ancestral RNA, about 88-base-long, is named "ur-RNA". Its primary structure was inferred by Hennig's (1965, 1966) concept of ancestor-descendant relationship (Fig. 1, Fig. 2), because Hennig's theory seems to provide us with a very efficient method for inferring ancestral sequences in the comparison of "closely related" sequences of informational macromolecules as demonstrated in the analysis of mammalian alpha-crystallin evolution by de Jong et al. (1977, 1985). The inferrence of ur-RNA sequence was made in each base position. The topology of evolutionary tree shown in Fig. 2 was assumed, since 5S rRNA can be considered to have evolved directly from a primitive tRNA (which we denote here by "proto-tRNA") by tandem duplication. In case 2 (or 3) in Fig. 2, for an example, base B in 5S rRNA (or tRNA) can be considered as to be "apomorphic" (derivative) character, and base A shared by tRNA (or 5S rRNA) and LTSV-A RNA 2 virusoid can be considered as to be "plesiomorphic" (primitive) character. Therefore, both ancestors at node 1 (proto-tRNA) and at node 2 (ur-RNA) can be predicted as to have had base A in the given position. Bases in other cases were similarly predicted.

Finally reconstituted primary sequence of the ur-RNA is given in Fig. 1 together with possible base-pairing secondary structure. This clearly concludes that ur-RNA is highly or nearly completely base-paired rod-like circular or non-circular RNA. This molecular structure and its similarity to virusoids strongly suggest that ur-RNA was, most probably, a self-replicating or auto-catalytic RO with no peptide-encoding informations, and replicated probably by the rolling circle model. Accordingly, it is most plausible that about 88-base-long, non-circular or circular, tRNA-like ancestor ("proto-tRNA"), with or without amino acid(Trp ?)-anchored structure, first evolved from a ur-RNA RO, and 5S rRNA thereafter evolved by tandem duplication of the proto-tRNA(s) or some primitive tRNA(s), and that proto-tRNA later speciated to generate the contemporary varieties of tRNA species. $tRNA^{Trp}$ seems to be the most primitive of contemporary tRNAs. Therefore, proto-tRNA, whether it self-replicated or not, must have been closely similar to the EC $tRNA^{Trp}$ in primary structure.

On the other hand, it remains to be important to answer the question whether the ur-RNA was a circular RNA as found in virusoids, or Trp(or anino acid)-anckored, non-circular RNA as in tryptophanyl(or aminoacyl)-tRNA. The possible evolutionary oldness of Trp-anckored tRNA might be a result from the similarity in chemical structure between tryptophan and nucleotide (base residue), by which the Trp in Trp-anckored ur-RNA could have easily interacted with (other) RNA. The oldness of $tRNA^{Trp}$ concluded as above seems to be very reasonable since the amno acid sequence of tryptophanyl tRNA-synthetase conserves a most primitive feature as compared with other aminoacyl tRNA synthetase sequences (Ohnishi, 1986c). This logical relationship needs to be further analyzed.

2.3. Speculations on the secondary structures of ur-RNA and its homologues

In the alignment of Fig. 1, Watson-Crick type U:A and G:C base pairs are indicated irrespective to the problem whether or not these base-pairs actually exist. The base pairs in B and C stems in the actual model of the secondary structure of 5S rRNA (Erdmann and Wolters, 1986), and those in the acceptor stem of the so-called secondary "clover-leaf model" of tRNA (Spinzel et al., 1985), are both conserved in the alignment in Fig. 1. On the contrary, possible or latently possible feature of highly base-paired rod-like secondary structure can be theoretically inferred for the EC tRNA as shown in Fig. 1, although it basically differs from the currently accepted clover-leaf model. Such highly base-paired rod-like model were found to be possible only for $tRNA^{Trp}$ (EC and BST), and not for other tRNA species in EC (Ohnishi, 1989a). Moreover, from the aspect of base-paiting structures, the EC $tRNA^{Trp}$ has a characteristic primary structure capable of forming the widely accepted clover-leaf secondary structure. Accordingly, there exists a

	case 1	2	3	4	5
++++++++ 5S RNA (BST) (1-76)	base A	<u>B</u>	A	A	A (B)
+++[1]++++++ $tRNA^{Trp}$ (EC)	A	A	<u>B</u>	A (B)	A
++++[2]					
++++++++++++ LTSV RNA 2 (virusoid)	A	A	A	C	C
Node [1] (proto-tRNA)	A	A	A	A (A or B)	A (A or B)
Node [2] (ur-RNA)	A	A	A	A or C (A,B or C)	A or C (A,B or C)

Fig. 2. Prediction of ur-RNA sequence. Bases underlined indicate apomorphic residues. Ancestral bases were inferred by Hennig's concept of ancestor-descendant relationship (Hennig, 1965). The inferrence of sequence was made in each base position. The topology of evolutionary tree shown here was assumed, since 5S rRNA can be considered to have evolved directly from tRNA by tandem duplication. In case 1 (or 2), base B in 5S rRNA (or tRNA) can be considered to be "apomorphic" (derivative) character, and base A shared by tRNA (or 5S rRNA) and virusoid can be considered to be "plesiomorphic" (primitive) character. Therefore, both ancestors at node 1 and at node 2 can be inferred as to have base A in the given position. Bases in other cases were similarly predicted.

[Ia] 2n-base plus-strand RNA with complete complementary symmetry (hairpin-like, completely base-paired structure)

$$\begin{array}{llll} 5'(1) & A_1\ A_2\ A_3\ A_4 \cdots\cdots A_{n-1}\ A_n & & (n) \\ & \ :\ \ \ :\ \ \ :\ \ \ : \qquad\qquad\quad :\ \ \ \ : &) & \\ 3'(2n) & B_1\ B_2\ B_3\ B_4 \cdots\cdots B_{n-1}\ B_n & & (n+1) \end{array}$$

[Ib] The same plus-strand RNA as [Ia]

$$5'(1) \quad A_1\ A_2\ A_3\ A_4 \cdots A_{n-1}\ A_n\ B_n\ B_{n+1} \cdots B_4\ B_3\ B_2\ B_1 \quad 3'(2n)$$

[II] 2n-base minus-strand RNA synthesis on plus-strand RNA template

$$\begin{array}{lll} 5'(1) & A_1\ A_2\ A_3\ A_4 \cdots A_{n-1}\ A_n\ B_n\ B_{n-1} \cdots B_4\ B_3\ B_2\ B_1 & 3'(2n) \\ & \qquad\qquad\qquad\quad :\ \ :\ \ : \qquad\quad :\ \ :\ \ :\ \ : & \\ 3' \xleftarrow[\text{RNA synthesis}]{} & B_n\ A_n\ A_{n-1} \cdots A_4\ A_3\ A_2\ A_1 & 5'\ (1) \end{array}$$

[III] 2n-base minus-strand RNA synthesized on plus-strand RNA template

$$3'(2n) \quad B_1\ B_2\ B_3\ B_4 \cdots B_{n-1}\ B_n\ A_n\ A_{n-1} \cdots A_4\ A_3\ A_2\ A_1 \quad 5'\ (1)$$

[IV] Palindrome symmetry of plus-strand/minus-strand base-paired complex

$$\begin{array}{lll} 5'(1) & A_1\ A_2\ A_3\ A_4 \cdots A_{n-1}\ A_n\ B_n\ B_{n-1} \cdots B_4\ B_3\ B_2\ B_1 & 3'(2n) \\ & :\ \ :\ \ :\ \ : \qquad :\ \ :\ \ : \qquad :\ \ :\ \ :\ \ : & \\ 3'(2n) & B_1\ B_2\ B_3\ B_4 \cdots B_{n-1}\ B_n\ A_n\ A_{n-1} \cdots A_4\ A_3\ A_2\ A_1 & 5'\ (1) \end{array}$$

Fig. 3. Complementary symmetry and palindrome symmetry in completely base-pared 2n-base plus-strand and minus-strand RNAs [Complementary symmetry model]. A_i and B_i is a base pair of U & A, A & U, G & C, or C & G. Colons (:) indicate Watson-Crick type base-paring by hydrogen bond. ")" denotes continuity of sequence. In this model, the minus-strand synthesized (III) on plus-strand is a complete copy (replica) of the original plus-strand. See text for details.

some possibility that $tRNA^{Trp}$ might have a rod-like secondary structure in some (or special) condition(s) other than those conditions in which it retains the widely accepted clover-leaf structure. If this would be the actual case, this possible rod-like structure would be a more primitive secondary structure than the clover-leaf one, and would be a result of evolutionary conservation from the old age of ur-RNA. This problem must be answered by direct structural analysis of $tRNA^{Trp}$, and the answer could not be inferred from the structure of other tRNAs such as $tRNA^{Phe}$.

These results in the secondary structures of ur-RNA, tRNAs and 5S rRNA, reveal a remarkable tendancy in which the secondary structure of these RNAs evolved from primitive, complementarily more (or highly) symmetric rod-like structure rich in Watson-Crick base-paires, to the advanced or specialized, non-rod-like (such as clover-leaf-typed) asymmetrical secondary structure. Evolution and diversification of ur-RNA thus occurred through breaking of complementary symmetry (namely, Watson-Crick Base-paired complementary feature) of secondary structure. We can not conclusively say whether or not complete complementarity was retained in the structure of ur-RNA, it must be noted that an RNA with complete complementarity would easily reproduce its progeny RNA identical to itself, if the progeny RNA could be synthesized by Watson-Crick hydrogen-bonding rule using the original RNA as a template.

A theretical model shown in Fig. 3, here named as "complementary symmetry model", is therefore worth to be examined, in which a 2n-base-long RNA with complete base-complementarity (Fig. 3, Ia and Ib) can easily reproduce an exact replica RNA by direct RNA synthesis on the original RNA template. That is to say, the newly synthesized (-)strand is an exact replica of the original (+)strand RNA (Compare III and Ib in Fig. 3). The plus-minus strand mechanism (which is found in MS2 RNA phage replication) is not needed for replication of RNA in this model. In the case of this model, completely base-baired rod-like structure basically similar to virusoids can be easily achieved as a plus-strand/minus-strand complex (= a two-molecule-complex) as found in Fig. 3-IV. It must, however, be noted that virusoids do not show such a "complete" base-complementarity.

The predicted secondary structure of ur-RNA shown in Fig. 1 is also rather asymmetrical in base-complementarity, even if it is "more" symmetrical than that of tRNA. it is, therefore, reasonable to conclude that the ur-RNA replicated most probably by some mechanism of plus-minus strand method more or less similar to the rolling-circle model (Saenger, 1987) in virusoids and viroids, or else, to MS2 RNA phage replication mechanism. By whichever plus-minus mechanism the ur-RNA replicated, the ur-RNA molecules themself must have behaved (or functioned) as ribozymes catalyzing their own or mutual replications. The close similarity of the ur-RNA with virusoid (= a ribozyme) is the most plausible reason for this speculative conclusion.

Fig. 4. (Next page) Alignment of 16S and 18S rRNAs with tRNAs and tRNA genes. EC = E. coli, BST = Bacilus stearothermophilus, PV = Phaseolus vulgaris, SC = Saccharomyces cerevisiae, CP = Cyanophora paradoxa. Chl. = chloroplast, Cyanl. = cyanelle. Base matches to rrnC $tRNA^{Trp}$ gene (EC), 16S, 18S, and 5S rRNA are indicated by "x", "o", "+", and ":", respectively. Modified bases in mature tRNAs are over-printed by slash (/). Spacer sequences in tRNA genes are written in lower case letters. A class I intron immediately downstream to position +35 of CP cyanelle trnL gene (indicated by "%") is omitted in this alignment. "###....###" indicates sequence segments in 16S and 18S rRNAs found, by Nazarea et al.(1985), to be homologous to $tRNA^{Val}$(GAC)(EC). Underlines indicate tri-peptides and longer peptides identical between tRNA (or its gene), small subunit rRNA (16S or 18S), and 5S rRNA(BST). Localizations of repeating tRNA-like sequence units in rRNAs are indicated by <=== ..... ===>. Align.I and Align.II are two different alignments of rrnC sequence against small subunit rRNAs. Data are from: $tRNA^{Ile}$ = Yarus & Barell, 1971; $tRNA^{Val}$ = Yaniv & Barell, 1971; $tRNA^{Leu}$(PV Chl.) = Osorio-Almeida, 1980; trnL, tRNA-Leu(CP Cyanl.) = Evrard, 1988; rrnC $tRNA^{Trp}$ = Yong, 1979; 16S rRNA (rrnB) = Huysman & Wachter, 1986, Brosius et al., 1981; 18S rRNA = Nairn & Ferl (1988). (Revised from Ohnishi, 1989b.)

```
tRNA^Ile(GAU) (EC)            5' (1) AGGCUUGUAGCUCAGGUGGUUAGAGC-GCACCCUG AUAAGGUGAGGCCCGCGGUUC-AAUCCGCGUUGGCCCA-CCA (74) 3'
 vs rrnC tRNA^Trp (x)                xxx   xxxx xxx  xxxx     xx  x   xx     xx     x  x   x  xxxx x xx      x xxx  xxx  % match(m/n)  Pnuc(m,n)
 vs 16S (o)                          o o  ooo o oooo o oo                 o  o o o    o       o  o o  o              ooo  34.7%(25/72)   .042
 vs 18S (+)                          + +   ++  +++++ ++++               +++  + +                   + +        +   +    +  31.5%(23/73)   .13
 vs 5S  (:)                            :: ::  :    : ::       :   :::::                :    ::::   ::  :   : :        :  35.7%(25/70)   .030
tRNA^Val(GAC) (EC)            5' (1) GCGUCCGUAGCUCAGUUGGUUAGAGCACCACCUUG ACAUGGUGGGGUCGGUGGUUCGAGUCCACUCGGACGCA-CCA (76) 3'
 vs rrnC tRNA^Trp (x)                 x  xxxxx xxx xxxxx     xxxxx    x     x    xxx xx xx  xxxxxxxx    xx x x x  xxx  60.8%(45/74)  .76 X 10^-10*
 vs 16S (o)                           o o oo o oooo o oo       o       o o     oo    o       o  o o  o   oo     o   ooo  37.8%(28/74)   .0099*
 vs 18S (+)                           + ++++  ++++++++++       +        ++      ++ +            ++   +    +  ++  +    +  38.7%(29/75)   .0062*
 vs 5S  (:)                              ::  :     ::    ::  : ::::    :       :      ::::   ::  :   : ::       :  34.7%(25/72)   .042

tRNA^Leu(UAA) (PV, Chl.)      5'(1) GGGGAUAUGGCGAAAUUGGUAGACGCUACGGACUU AAAAUCCGUCGACUU/ (50)
vs rrnC tRNA^Trp (x)                 xxx   x x   xxxxxxxxx xx  xxx xx   xx xxx    x
                                                                     (51)  /AAUAAAUCAUGAGGGUUCAAGUCCC-UCUAUCCCCACCA (88) 3'
                                                                             x    x  xx x xxxx xxxx x xx   xxx  xxx
trnL,tRNA^Leu(UAA)(CP,Cyanl.)5'(1) GGGGATGTGGTGGAATTGGTAGACGCAACGGACTT%AAAATCCGTCGATTG/  (50)
 vs rrnC tRNA^Trp (x)                xxx  xx xx  xxxxxxxxx xxx xxx xx   xxx xxx    xxx
                                                                     (51)  /TATAGATCGTGAGGGTTCAAGTCCC-TCCGTCCCC-Atttaatttt-tctattttcagt
                                                                                x x  xx x xxxx xxxx x xxxx xxx         xxx  x  xx          xx
rrnC (EC),tRNA^Trp(Align.I) 5'(+1) AGGGGCGTAGTTCAATTGGTAGA-GCACCGGTCTC CAAAACCGGGTGTTGGGAGTTCGAGTCTC-TCCGCCCCTGCCAgaaatcat-ccttagcgaaagc (+97)
 vs 16S (o)                          o oo  oo o ooo  o oo      o    oo       oo             ooo  o o  oo  ooo     oooooo   o                 o
 vs 18S (+)                          + ++ +++    +++ +++++ ++    +    ++++  +  ++  + +       ++     ++  ++   ++  +        ++   +       +      +
 vs 5S  (:)                                :       :  :: ::: : : :    :       : :  :       : :::     ::::  :::: ::  ::     :   ::      ::

16S rRNA (EC)              (1055) /AUGGCUGUCG-UCAGCUCGUGUUGUGAAAUGUUGG GUUAAGUCCCGCA-ACGAGCGCAACCCUUAUCCUUUGUUGCCAGCGGU-------------CCGG (1139)
 vs 18S (+)                        +++++ ++   ++++ + +++  ++++  +++  +  ++++ + +   + ++++ ++   ++ ++++ +++            ++ +++              + ++
18S rRNA (yeast, SC)       (1271) /AUGGCCGUUUCUCAGUUGGUGGAGUGAUUUGUCUG CUUAAUUGCGAUA-ACGAACGAGAC-CUUAACCUACUAA-AUAGUGGU-GC-UAGCAUUUGCUGG (1365)
homology to tRNA^Val (#)                 #################
         (Nazarea & al.,1985)

rrnC (EC),tRNA^Trp (Align.II)                                                                      (-4) /tattAGGGGCGTAGT-TCAATTGG  (+19)
 vs 16S (o)                                                                                                  o   o            oo
 vs 18S (+)                                                                                                  + + +  ++ +++   +   ++++
 vs 5S  (:)                                                                     AA                           :: : :::::  ::    :  ::
5S rRNA(BST)               5'(1) --CCUAGUGACAAUAGCGG-AGAGGAAACACCCGU UCCCAUCCCGAAC-ACGGGUUAAGCUCUCCAGCGCCGAUGGUAGUUGGGGC-CAGC-GCCCCUGC   (96)
 vs 16S (o)                            oo        o    o  oo      o      o  oo    ooo o    o o oo    o    o oo  oo  o               o o
 vs 18S (+)                            ++   +      ++  +++  +       +      ++  +++   +++     +  + ++ ++ +   + +   ++++ +  ++  +++        +++
 tRNA-like repeats in rRNAs      <=========================== tRNA-like repeat 1 ===========================>  <==================
......................................................................................................................................
trnL, tRNA^Leu (UAA) (CP,Cyanl.)  ta------tcatttaattaaa-aaaa---tctaaatttttaaattgtt/
 vs rrnC tRNA^Trp                 xx      x  xxxx  x xx x xx   xxx        xxxxx  xx
rrnC (EC), tRNA^Trp (Align.I)(+98) taaggattttttttatctgaa-ataaccctct--ccgaagtaaatccttc-taccggca/ (+152)
 vs 16S                              ooo  o          oo     o       o o oo oo        o  oo oo      40.0%(53/136) Pnuc = .23 X 10^-3*
 vs 18S                            + +      ++ ++     ++  ++      ++    ++++ + ++ +      +     ++      42.6%(63/148) Pnuc = .23 x 10^-5*
 vs 5S                              :   :     :  ::     :                                          36.3%(41/113) Pnuc = .0050*

16S rRNA (EC)              (1140) CCGGGAACUCAAAGGA--GACUGCCAGUGAUAAACUGGAGGAAGGUGGGGAUGACGUCAAGUC-AUCAUGGCCCUUA---CGACCAG/ (1222) Pnuc
 vs 18S                                 +++       ++  ++++  + ++    ++ + +  +++++ + + +   +    ++ +++   + +++ ++++++    ++  + +         = .11X10^-19*
18S rRNA (SC)              (1366) UUAUCCACUUCUUAGAGGGACUAUCGGUUUCAAGCCGAUGGAAGUUUGAGGCAAUAACAGGUCUGUGAUG-CCCUUAGAACGUUCUG/ (1450) 59.5%(97/163)

rrnC (EC),tRNA^Trp(Align.II) (+20) TAGAGCACCGGTCTCCAAAACCGGGTGTTGGGAGTTCGAGTCT-TC-TCCGCCCCTGCCAgaa-atcat---ccttag--cga-aag/ (+96)          Pnuc
 vs 16S                              o o oo              oo o   oo     o  o ooo                    o  o oo  ooooo    ooooo   ooo  oo 43.5%(37/85) .00015*
 vs 18S                            +    +++    +         ++     +++     +       +      +       ++      + +     + ++   ++++++  ++     + 40.8%(40/98) .00042*
 vs 5S                              :::: :         : : :   :                                                                          51.2%(22/43) .00020*
5S rRNA (BST)                 (97) AAGAGUAGGUCGUUGCUAGGC (117) 3'
 vs 16S                             o o o        o    o o       32.7%(33/101)  Pnuc = .051
 vs 18S                                 +   ++ + +    + +       41.1%(46/112)  Pnuc = .14 X 10^-3*
tRNA-like repeats in rRNAs         ========= tRNA-like repeat 2 ==============================>
```

The ur-RNA of ca. 88 bases less probably replicated by complementary symmetry model (in Fig. 3-II), but it is yet worth examining a hypothesis that, at a very early stage of RNA evolution, some shorter RNA ($2n < 88$) with complete base-pairing complementarity might have emerged and replicated by the complementary symmetry model. This problem will be further discussed later in section 10.

3. Homology of tRNA genes and their 3'-flanks with rRNAs

Before contemporary cell machinery began to work, each ur-RNA was an individual of ribo-organism (RO) capable of replicating by the catalytic activity of its own self or of other ribo-organismic individual(s). Thus, proto-tRNA and therefrom derived tRNAs can also be considered to be ROs, even if they partially or completely lost their capability of self-replication. The deficiencies in self-replication must have occurred throughout adaptive evolution of co-operative behaviours of these ROs. This phenomenon seems to paralell to the fact that chloroplasts and mitochondria in higher plants and animals cannot self-replicate outside the host cell as a result of highly inter-dependent symbiosis with the host.

By tandem duplication of proto-tRNA or a very primitive tRNA, an RNA with one and a half unit length of tRNA evolved as a 5S rRNA. 16S and 23S rRNAs might also have evolved by repeated duplications of these tRNA-like RNA units (tRNAs and/or rRNAs) or by linear arrangement of them, because many sequence segments in EC 16S rRNA and yeast 18S rRNAs show significant homologies to tRNA(s) (Bloch et al., 1983; Nazarea et al., 1985). A most remarkable homology hitherto reported is in residues 5-20 of E. coli $tRNA^{Val}$ (with anti-codon GAC) with residues 1059-1073 (16S) and 1275-1290 (18S) of small subunit rRNAs (SS-rRNAs) (Nazarea et al., 1985). Based on Nazarea et al.'s finding of this homology and on the alignment in Fig. 1, tRNAs and their genes as well as 5S rRNA were compared with these SS-rRNAs. The resulting alignment is shown in Fig. 2. In the alignment I (Fig. 4) of the bases +1 - +152 of EC rrnC operon including $tRNA^{Trp}$ gene and its 3'-flank ("+1" corresponds to 3'-terminus of mature $tRNA^{Trp}$), with bases 1055-1206 (EC) and 1271-1423 (yeast) of SS-rRNAs, giving 40 and 43% base matches with strong significance levels, $P_{nuc} = .23 \times 10^{-3}$ and $.23 \times 10^{-5}$, respectively. Spacer sequence of rrnC operon flanking 3'-side of $tRNA^{Trp}$ gene was found to show considerable homology to yeast 18S rRNA. The $tRNA^{Trp}$ gene (-4 - +96) was further aligned with bases 1129-1222 of EC 16S rRNA ("Alignment II" in Fig. 4) giving 44% match and $P_{nuc} = .15 \times 10^{-3}$. 5S rRNA (BST)(1-117) also shows 41% match and $P_{nuc} = .14 \times 10^{-3}$ in comparison with bases 1271-1386 of 18S rRNA.

The sequence region of SS-RNAs shown in Fig. 4 is thus concluded to have at least two tRNA-like complete sequence units tandemly arranged. A dinucleotide (1356-1357 in yeast 18S RNA) exists between these two units. Considering that there plausibly exist many other tRNA-like sequences in SS-rRNA (Bloch et al., 1983), SS-rRNAs must have evolved as "multi-tRNA ROs", which we shall further discuss in the next section. 5S-rRNA is therefore a di-tRNA RO.

Furtheremore, it must be noted that the E. coli 16S RNA segment (bases 1055-1206) with di-tRNA structure corresponds to the helical domain proposed by Murgóra et al. (1989) for termination at all three termination codons, UGA, UAA and UAG.

4. MULTI-tRNA RIBO-ORGANISMS AS SYMBIOTIC COMPLEXES OF DIFFERENT tRNA (or PROTO-tRNA) RIBO-ORGANISMIC INDIVIDUALS

In EC, tRNAs and rRNAs are derived from larger short-lived precursor

RNAs that ribonucleases (RNases) break down into mature tRNA(s) and/or rRNAs. Very few specific DNA regions encode 30 to 40 different pre-tRNAs, whereas 30S pre-rRNA transcripts including 23S, 16S and 5S rRNAs in that order are encoded by seven rrn operons (rrnA, rrnB, ... , rrnH). Some tRNA genes are embedded in rrn operons (Watson et al., 1987). Fig. 4 strongly suggests that 16S rRNA is a direct homologue to some pre-tRNA(s) which also carries many tRNA individuals and some spacers. Thus we can postulate that pre-tRNAs and pre-rRNAs might have evolved by repeated duplications or symbiotic linear arrangements of original proto-tRNAs and/or therefrom derived RNA individuals. Accordingly, rRNAs are ROs which evolved as linearly arranged symbiotic complexes of unit-length tRNA ROs. Messenger RNAs (mRNAs), pre-tRNAs and pre-rRNAs in contemporary cells are all transcripts generated by RNA polymerases, suggesting a possibility that mRNAs might also be ROs comparable to pre-tRNA or the like.

We can therefore propose a hypothesis that;

> [H 1]: Genetic apparatus including various types of contemporary ROs (tRNAs, rRNAs and mRNAs) evolved as a resulting evolutionary product of adaptive, mutually co-operative symbiosis, or in other words, of a co-operative work adaptively achieved by the population of these ribo-organismic individuals.

In [H1], contemporary protein-synthesizing mechanism can be considered as a "culture" of the co-operative "society" of these ROs, or as a kind of "meme" in the society if we use Dawkin's (1976, 1982) terminology because this mechanism can replicate by using the ribo-organismic society itself. This society consists of mutually co-operating ribo-organismic individuals of different species, and is, therefore, a kind of evolutionarily stable "eco-system" or "community". DNA genes and proteins are "tools" or "machines" in this culture, and therefore are considered to be cultural products of the society. These "cultural products" including various DNA genes and their direct and indirect products (such as proteins and other phenotypes) must have provided the society members (ROs) with higher Darwinian fitness to cause advantages in their indirect replication using the cell-machinery, and have therefore evolved by natural selection.

Accordingly, we can find a hierarchy of individuality consisting of a lower level of ribo-organismic individuality and an upper level of (haploid or prokaryotic) cell-individuality. Each of these two levels of individuality ever lived or presently lives as independent individuals. Evolution of hierarchical structure in individuality and culture will be further discussed in the later section.

For further examining this symbiotic hypothesis, we must answer to the question whether or not protein-encoding genetic informations also evolved as a co-operative work (culture) of these ROs. For answering this, we need to compare gene sequences encoding most primitive enzymes with LTSV-A RNA 2 (virusoid) and EC rrnC $tRNA^{Trp}$ gene as well as with other related and possibly related RNAs such as M1 RNA (RNA component) of ribonuclease (RNase) P. For this purpose, the "primodial gene theory" (Ohnishi, 1986a,b) will be reviewed and re-considered in the next section, and in the later section, comparisons of the virusoid and rrnC $tRNA^{Trp}$ gene with some primitive enzyme gene segments derived from a most primitive primodial gene will be further discussed and examined from a viewpoint of ribo-organsmic evolution.

5. A MOST PRIMITIVE PRIMODIAL GENE ENCODING ca. 47-aa-PEPTIDE

5.1. Moore and Goodman's alignment statistic for evaluating aa homology

Moore and Goodman's alignment statistic, here denoted by $P_{aa}(M,N)$, is useful for evaluating homology levels in aligned two aa sequences (Moore and Goodman, 1977). $P_{aa}(M,N)$ is defined as the probability that an N-aa-long

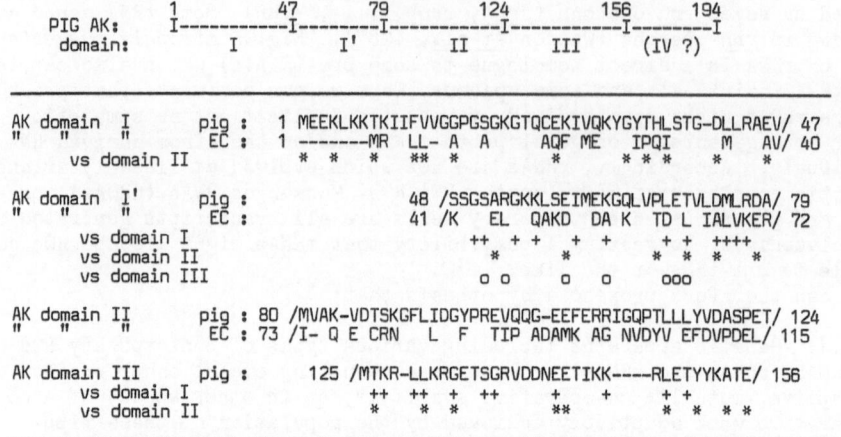

```
        1          47       79         124         156       194
PIG AK: I-----------I---------I-----------I-----------I---------I
domain:       I          I'        II         III       (IV ?)
```

```
AK domain I    pig :  1 MEEKLKKTKIIFVVGGPGSGKGTQCEKIVQKYGYTHLSTG-DLLRAEV/ 47
"   "     "     EC :  1 -------MR LL- A  A     AQF ME    IPQI     M   AV/ 40
   vs domain II         *  *   *  **  *        *      * * *    *    *

AK domain I'   pig :              48 /SSGSARGKKLSEIMEKGQLVPLETVLDMLRDA/ 79
"   "     "     EC :              41 /K   EL  QAKD  DA K  TD L IALVKER/ 72
    vs domain I                         ++  + +  +   +    + +  ++++
    vs domain II                          *      *        *  *  *   *
    vs domain III                        o o     o   o     ooo

AK domain II   pig : 80 /MVAK-VDTSKGFLIDGYPREVQQG-EEFERRIGQPTLLLYVDASPET/ 124
"   "     "     EC : 73 /I- Q E CRN   L  F  TIP ADAMK AG NVDYV EFDVPDEL/ 115

AK domain III  pig :      125 /MTKR-LLKRGETSGRVDDNEETIKK----RLETYYKATE/ 156
    vs domain I               ++  ++  +  ++       +      +      + +
    vs domain II              *    *  *           **            *  *  * *
```

Fig. 5. Internal homology units (domains) of adenylate kinase (AK). Amino acid (aa) sequences of the N-terminal three domains of pig skeletal muscle AK (myokinase) (194 aa's) and E. coli (EC) AK (214 aa's) and of pig AK III domain are aligned with one another. In EC AK, only those aa residues different from pig AK are shown. "-" indicates gap. Residues identical to those in domain I, II and III are indicated by "+", "*" and "o", respectively. Domain AK I' is shorter than other ca. 47-aa domains. The pig AK IV sequence segment also shows weak similarities to other domains. Data are from: pig AK = Heil et al., 1974; EC AK = deduced from adk gene (Brune et al., 1985).

```
                      1             47            95            139
EPSILON  EC:          I-------------I-------------I-------------I
                                     ::::::::::::: +++++++++++++
       domain:               I             II           III


                             1     23            69            112         149
GAMMA    EC:                 I------I-------------I-------------I-----------I*
                                     ::::::::::::: +++++++++++++
       domain:                 I(?)        II           III         ( IV? )


                      150          192           242           287
                       *I------------I-------------I------------I
                                                    ++++++++++++
       domain:                V            VI           VII
```

Fig. 6. Domain structures of epsilon (139 aa's) and gamma (287 aa's) subunits in E. coli F_1-ATPase. Numbers denote residue-number. In gamma subunit, whether the sequence segments I and IV are also homologous to other domains is unknown. As shown in Fig. 8, other domains in both subunits are homologous with one and with AK domains. Gamma-VII is highly homologous to gamma III and epsilon III. Gamma II is highly homologous to epsilon II. Based on Ohnishi (1986a,b).

alignment of two aa sequences deduced from randomly produced nucleotide sequences gives MMD ≤ M where MMD denotes Fitch and Margoliash' (1967) minimum mutation distance between the two sequences. If an alignment of two actual aa sequences gives MMD = M, then $P_{aa}(M,N)$ can be used as an statistical index for evaluating the degree of homology level. P_{aa} is based on multinominal distribution as clearly demonstrated by Vogel (1978), and can be easily computed by a FORTRAN program "PROBAA" made by K. O. (Ohnishi, 1984a).

5.2. Primodial gene theory (Ohnishi, 1986a, b)

Adenylate kinase (AK) and F_1-ATPase are key enzymes of energy-yielding systems in every of contemporary living organisms. The primodial gene theory (Ohnishi, 1986a, b) postulated that a most primitive primodial gene first evolved as to encode a ca. 47-aa-peptide co-ancestral to the domains of AK, F_1-ATPase epsilon and gamma subunits, and EC ribosomal protein (r-protein) L34 (46-aa-long).

As shown in Fig. 5, both of pig skeletal muscle AK (alternative name: myokinase) and EC AK consist of several internal repeating units of ca. 47 aa's. Domain I (pig AK, aa residues 1-47) and domain I' (48-79) are highly homologous with each other (31% aa match and P_{aa} = $.14 \times 10^{-4}$ between pig AK I and pig AK I'), and domains II (80-124) and III (125-156) is also significantly homologous to the domain I (20% aa match and P_{aa} = $.64 \times 10^{-2}$ between pig AK I and pig AK II). The remaining C-terminal region (segment IV in Fig. 5) (157-194) also seems to show some similarity to other domains. The AK's are thus concluded to be composed of four or five homologous domains which are basically ca. 47-aa-long repeating units (Ohnishi, 1986a, b).

Fig. 6 further illustrates domain structures of epsilon (139 aa's) and

Table 2. Representative cases of statistically significant homologies in the alignments of ca. 47-aa domains of house-keeping enzymes shown in Fig. 7 and Fig. 8. #

Comparison of domains	aa sequence % match	aa sequence P_{aa}	base sequence % match	base sequence P_{nuc}
[epsilon II-like domains]				
EC epsilon I vs EC eps. II	15%*	(.20)	38%*	.00035
EC gamma II vs Chl. eps. II	23*	.0064	39*	.00019
[epsilon III-like domains]				
EC gamma III vs EC eps. III	20*	(.012)	39*	.00040
EC gamma VII vs EC eps. III	21*	.0064	38*	.00060
EC gamma V vs EC gamma VII	14	.0031	41*	.000097
EC r-protein L34 vs EC eps.III	14	(.036)	--	------
EC r-protein L34 vs Chl. eps.III	11	.0064	--	------
[EC Trp-RS, NRU2 domain]				
Trp-RS NRU2 vs EC eps. II	17*	(.021)	42*	.000021
Trp-RS NRU2 vs EC eps. III	19*	.0069	35	.0088
[AK domains]				
pig AK II vs EC eps. III	19*	.0079	--	------
EC AK II vs Chl. eps. II	9	(.069)	36	.0026

P_{aa} and/or P_{nuc} values in the alignments of Fig. 7 and Fig. 8 are tabulaed. Asterisks(*) indicate aa match ≥ 15% and base match ≥ 38%, and further, non-significant values of P_{aa} and P_{nuc} are parenthesized. It must be noted that P_{nuc} is more sensitive than P_{aa} in detecting genuine homologies. Abbreviations used in this Table: Chl. = <u>Zea may</u> chloroplast; Trp-RS = tryptophanyl tRNA synthetase.

gamma (287 aa's) subunits of F_1-ATPase in EC. These subunits also consists of ca. 47-aa domains having aa sequences homologous to AK domains. Three domains in epsilon (named epsilon I, II, III), and at least six domains in gamma (named gamma I-III, V-VII) can be detected. Whether the region IV (residues 113-149) in gamma subunits is also homologous to other domainns is unknown. (Domain gamma I is also yet somewhat doubtful.)

Fig. 7 shows homology relationship among the gene sequence segments coding for these and related ca. 47-aa-domains of primitive enzymes including primitive AK domains, primitive domains of F_1-ATPase epsilon and gamma subunits, and a possibly related domain of tryptophanyl tRNA synthetase (Trp-RS) [the 2nd N-terminal repeating unit (NRU 2), see Ohnishi (1986) for details] and EC RNA polymerase (core enzyme) alpha subunit (C-terminal domain). P_{nuc} values are also given in the right side of Fig. 7.

The homology relationship among the aa sequences of representative domains of these and of r-protein L34 are given in Fig. 8, where the alignment is based on Fig. 7. Table 2 summarizes representative comparisons giving statistically significant levels of homology in aa and/or corresponding nucleotide sequences. It should be reminded that P_{nuc} is more sensitive than P_{aa} in detecting genuine homologies.

In the comparison between pig AK domain II with EC ATPase epsilon III domain, a significant homology of 19% aa match giving P_{aa} = .0079 was obtained. In the alignment of the EC L34 r-Protein with epsilon III domains of EC and Zea may chloroplast, weakly significant homology levels giving 14% aa match with P_{aa} = .036 (EC) and 11% with .0064 (chloroplast), respectively, were observed (Table 2).

Statistical analysis of homology relationships among the domains of ATPase epsilon and gamma subunits revealed not only that these domains are homologous to one another, but also that a two-domain peptide, epsilon II-III, is highly homologous to gamma II-III peptide, as illustrated in Fig. 6. This conclusion can be obtained from the considerably significant levels of homology between maize chloroplast epsilon II domain and EC gamma II domain (P_{aa} = .0064, P_{nuc} = .00019), and between EC epsilon III domain and EC gamma III domain (P_{nuc} = .00040). Moreover, AK II domain is more highly homologus to epsilon III domain (P_{aa} = .0079 between pig AK II and EC epsilon III) than to epsilon II domain(Table 2). Thus it is most plausible to conclude that the direct, the last ancestror of epsilon and gamma subunits was or had a two-domain-long (ca. 94-aa-long) enzyme or peptide segment closely related to the epsilon II-III peptide, and that the more ancient enzyme commonly ancestral to this ancestral enzyme and AK was a two-domain peptide (Fig. 9).

It can, therefore, be concluded that the ca. 47-aa domains of AK II,

Fig. 7. (Next page) Alignment of nucleotide and amino acid sequences of internal repeating units of the genes encoding F_1-ATPase epsilon and gamma subunits, adenylate kinase (AK), tryptophanyl tRNA synthetase (Trp-RS), and core enzyme subunit alpha. Roman numerals, I, II, III, et al. denote numbering of domains as defined in Figs. 5 and 6. EC = E. coli, BS = Bacillus stearothermophilus, Chl. = maize chloroplast, Trp-RS = Trp'yl tRNA synthetase. Nucleotide sequences and deduced aa sequences are aligned, and protein sequences of pig AK and Trp-RS are also aligned. Numerals at the left and right ends of each aa sequence denote numbering of aa residues, and define boundaries of each domain. "O" denotes stop codon. Non-coding nucleotide sequences are written in lower case letters. Base matches to epsilon II/I domains, to epsilon III, and to gamma domains are indicated by ":", "+", and "*", respectively. Base matches to AK domains and to Trp-RS/core enzyme are indicated by "=" and "o", respectively. On the right hand, percent base matches and matching probabilities (P_{nuc}) are given for each comparison. Aa's matched among AK domains are underlined. See Ohnishi (1986b) for references of original sequence data. (Slightly revised from Ohnishi, 1986b.)

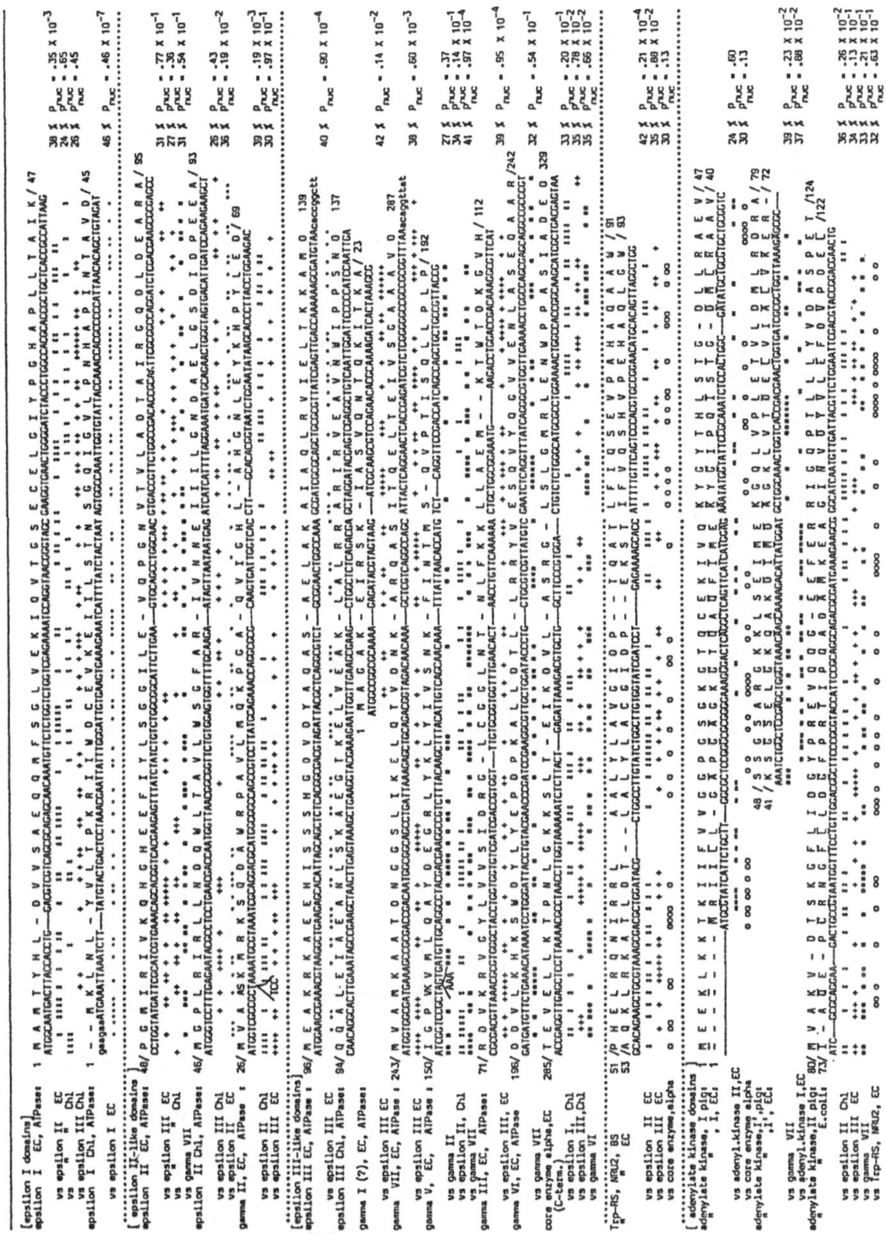

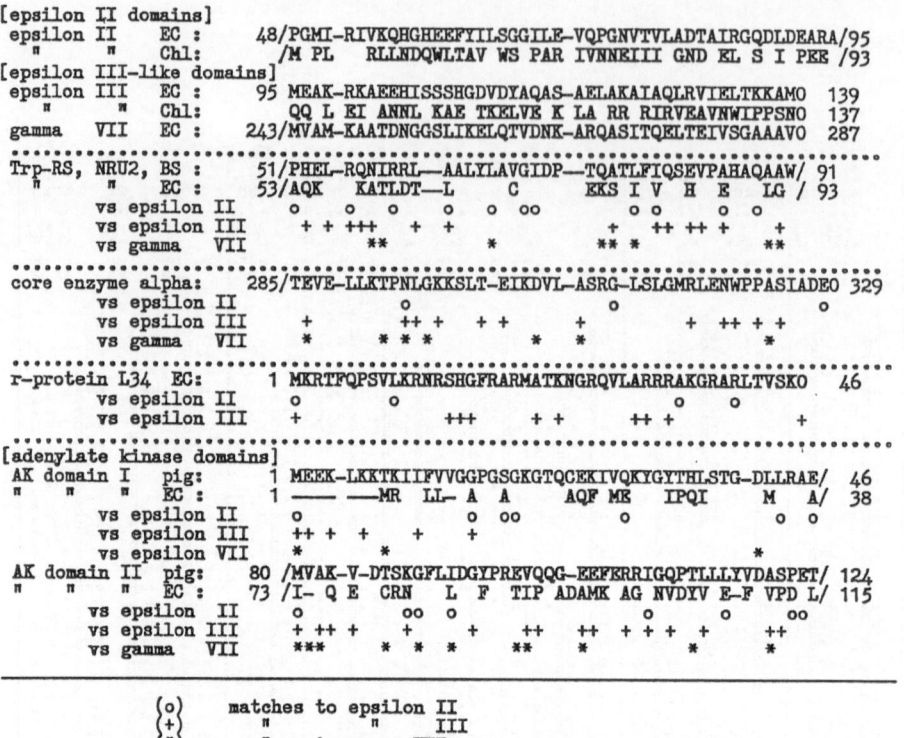

```
[epsilon II domains]
 epsilon II    EC :      48/PGMI-RIVKQHGHEEFYILSGGILE-VQPGNVTVLADTAIRGQDLDEARA/95
   "     "     Chl:        /M PL   RLLNDQWLTAV WS PAR IVNNEIII GND EL S I PEE /93
[epsilon III-like domains]
 epsilon III   EC :      95 MEAK-RKAEEHISSSHGDVDYAQAS-AELAKAIAQLRVIELTKKAMO  139
   "      "    Chl:         QQ L EI ANNL KAE TKELVE K LA RR RIRVEAVNWIPPSNO  137
 gamma   VII   EC :     243/MVAM-KAATDNGGSLIKELQTVDNK-ARQASITQELTEIVSGAAAVO  287
.................................................................................
 Trp-RS, NRU2, BS :      51/PHEL-RQNIRRL—AALYLAVGIDP—TQATLFIQSEVPAHAQAAW/ 91
   "     "     EC :      53/AQK   KATLDT—L        C       EKS I V  H  E   LG / 93
          vs epsilon II     o     o    o     o    o  oo           o o     o  o
          vs epsilon III     + + +++     +   +                 +    ++ ++ +    +
          vs gamma   VII          **             *             ** *            **
.................................................................................
 core enzyme alpha:     285/TEVE-LLKTPNLGKKSLT-EIKDVL-ASRG-LSLGMRLENWPPASIADEO 329
          vs epsilon II                o                   o                   o
          vs epsilon III     +         ++ +    + +         +         +   ++ + +
          vs gamma   VII     *        * * *          *     *               *
.................................................................................
 r-protein L34  EC:       1 MKRTFQPSVLKRNRSHGFRARMATKNGRQVLARRRAKGRARLTVSKO   46
          vs epsilon II     o          o                         o     o
          vs epsilon III    +               +++        + +      ++ +            +
.................................................................................
[adenylate kinase domains]
 AK domain I    pig:      1 MEEK-LKKTKIIFVVGGPGSGKGTQCEKIVQKYGYTHLSTG-DLLRAE/  46
 "     "   "    EC :      1 ——— ——MR  LL- A   A     AQF ME   IPQI      M   A/  38
          vs epsilon II     o                o  oo       o                o  o
          vs epsilon III    ++ +  +     +    +
          vs epsilon VII    *        *                               *
 AK domain II   pig:     80 /MVAK-V-DTSKGFLIDGYPREVQQG-EEFERRIGQPTLLLYVDASPET/ 124
 "     "   "    EC :     73 /I- Q E  CRN    L  F  TIP ADAMK AG NVDYV E-F VPD L/ 115
         vs epsilon  II     o          oo  o                 o       o      oo
         vs epsilon III     + ++ +      +      +       ++     ++   + + +   +   ++
         vs gamma   VII     ***        *   *   *        **     *          *      *

             (o)     matches to epsilon II
             (+)        "          "    III
             (*)        "     to gamma VII
```

Fig. 8. Alignment of amino acid sequences of representative domains of adenylate kinase (AK), F_1-ATPase epsilon and gamma subunits, tryptophanyl tRNA synthetase (Trp-RS), r-protein L34, and core enzyme subunit alpha. Aa sequence data of EC L34 is from Chen, 1976. See legend of Fig. 7 for abbreviations.

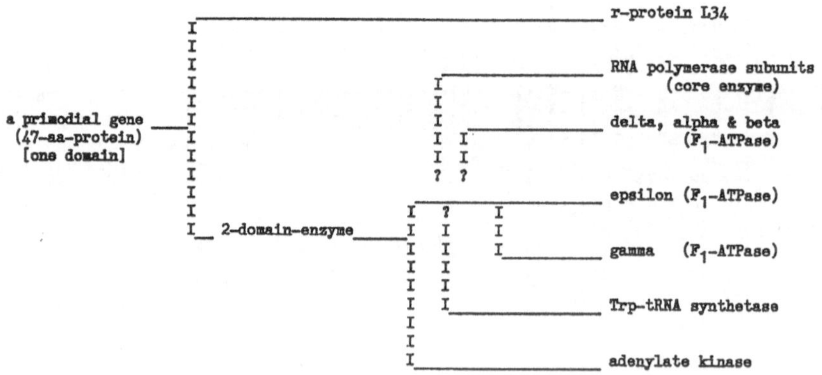

Fig. 9. Tentative phylogenetic tree of primitive house-keeping enzymes as evolved from a ca. 47-aa-long domain encoded by a most primitive primodial gene.

EC F_1-ATPase epsilon III and II have well conserved a feature of a most primitive sequence of primodial enzyme closely similar to L34. Whether the C-terminal sequence region (aa residues 285-329) in EC core enzyme alpha subunit (Fig. 8) is also a genuine homologue to these 47-aa domains is yet doubtful and remains to be further examined.

Accordingly, it is concluded that a primitive primodial gene encoding ca. 47-aa-long enzyme emerged in early life, and therafter evolved to have generated many house-keeping enzymes including AK, ATPase subunits and amino-acyl tRNA synthetases (ARS's), since F_1-ATPase alpha and beta subunits also seem to have internally repeating homology units related to AK domains (Ohnishi, unpublished).

It is worth noticing that L34 is the smallest r-protein in EC ribosome, and that L34 is exactly one-domain-long, meaning that L34 gene is most probably a direct descendant of the first primodial gene coding for the ca. 47-aa primodial protein. A most plausible evolutionary tree inferred from these results is given in Fig. 9.

5.3. Origin of MS2 replicase and mammalian DNA polymerase beta

Furthermore, the entire sequence of RNA replicase beta (aa's 1-544) in MS2 RNA phage has recently been found to be homologous to F_1-ATPase alpha (aa's 1-513) and beta (aa's 1-460) subunits as well as to mammalian DNA polymerase (DPase) beta (aa's 1-335) and terminal deoxyribonucleotidyl transferase (TDT) (aa's 1-508) and to T7 and sp02 phage DPases and EC Pol I DPase (Ohnishi, 1989c). The alignment of these nucleic acid replicases is given in Fig. 10, whose detailed statistical analysis will be publshed elsewhere. This alignment has been achieved by referring to Ollis et al.(1985) and Ohnishi(1987) for the homology between Pol I and T7 DPase, to Matsukage et al.(1987) for that between TDT and mammalian DPase beta, and to Ohnishi (1988) for a preliminary alignment.

Taking it into consideration that ATPase alpha and beta subunits seem to also have internal repeating units similar to the domains of AK and ATPase epsilon subunit (Ohnishi, unpublished data), the genes for these nucleic acid replicases almost undoubtedly evolved by duplications of the above-mentioned primordial gene. This strongly confirms that contemporary nucleic acid replicases/polymerases evolved directly from a primitive enzyme of energy-aquiring and/or -providing systems, which is the direct ancestor of contemporary F_1-ATPase alpha/beta subunit. The enzyme co-ancestral to F_1-ATPase alpha/beta and MS2 replicase/rat DPase beta was probably an NTPase which can cooperate with some enzyme or ribozyme to work as an RNA replicase.

6. SMALL RIBOSOME THEORY (Ohnishi, 1984b)

We thus have reached an important conclution that L34 is a most primitive protein, which well coincides with the previously proposed "small ribosome theory" (Ohnishi, 1984b).

The number of r-proteins range between 50-60 in bacterial ribosome and 70-80 in cytoplasmic ribosome of eukaryotes. EC r-proteins are 46 (L34) to 233 (L1) aa's long exept a longer protein of S1 (Wittmann et al., 1980).

There are many statiistically detectable (by P_{aa}) homology relationships among aa sequences of these r-proteins, and based thereupon, Ohnishi (1984b) tentatively classified 35 r-proteins into 10 different classes. The classification in Fig. 11 includes 36 r-proteins because homology can be found between L34 and L30. Thus these two and L32, L29 and L33 are homologous to one another, and members of class L33. This means that there must be much more genuine homology relationships among EC r-proteins which are statistically undetectable. Taking this fact in consideration, I concluded, in the paper of the small ribosome theory (Ohnishi, 1984b), as follows;

"most or all of contemporary r-proteins must have evolved by repeated gene duplications of very few (or only one) primitive ancestral r-protein

```
Pol I, EC             207 /GLGGLDTLYAEPEKIAGLSFRGAKTMAAKLEQNKEVAYLSYQ---LATIKTD--VELEL------------------TCEQLEVQQP-AAEELLGLfkky
sp02 DPase          1 MKTLSIDIETFL-SVDLLKAGV-YAYTEAPDFEIL/33
  "                                   34 /lfaYAFDDD-PVKIIDLAQGD--TLPHEVLVALTSSKVIKT-AYKA-----NFERTCIAKHFNLM--LLPAQWRCTA-
T7 DPase           1 mivsdieaNALLESVTKFHCGVIYDY--STAEYVSYRPS-D-F--GAYLDALEAEVARGGLIVFHNGHKYDVP-ALTK-----LAK---LQL--NRE-FHLPRE-NCID- 94
TdT, human         1 MDPPRASHLSPRKKRPRQTGALMASSPQDIKFQDLVVFILE-KKMGTTRRAFLMELARRKGFR-VENELSDSVTHIVA-----ENNSGSDVLEWLQ--AQKVQVSSQPE- 100
DPase beta, rat    1 MSKRKAPQETLNGGITDMLV> 20 (exon 1)
repl. MS2          1 MSKTTKKFNSLCIDLPRDLSLEIYQSIASVATGSGDPHSDD-FTAIAYLRDELLTKHPTLGS-GNDEATRRTL-AIAK-----LREANGDRGQINREGFLHDKS-LSW-- 99
ATPase alpha,EC    1 MQLNS-TEISELIKQRIAQFNVVSEAHNEGTIVSVSDGVIR-IHGLADCMQGEMISLPGNRYAIALNLERDSV-GAV-------VMGPYADLAEGMKVKCTGRILEVP--
   "  beta, EC     1 ------------------------MATGKIVQVIGAVVD-V-EFPQDAVPRVYDALEVQN-GNERLVLEVQQQLGGGIVRTIAMGSSDGLRRGLDVKQLEHPIEVP--
..............................................................................................................................
Pol I            289 EFKR-WTTADVEAGKWL-------------QAKGAKPAAKPQETSVADEAPEVTATVISYDNYVTI--LDEETLKAWI--AKLEKAP--VFAFDTETDSLDNISANLV 369
sp02 DPase           VH---ATTLG-Lpgn-l--------------dGVAK-ALKLSAQKDKAGKALIRYFSVPCK-PTKA--NGQRVRNLPE--HDPEKW---EKFKVY-CIDVEVERAIKN
T7 DPase           95 TLVLSRLI----HSN-L-------------KDTDMGI-----------------LRSGKLPGKR--FGSHALEAWG--YRLGEMKG-EYKDDFKRMLEEQ-GEEYV 155
TdT              101 LLDVSWLIECIGAGKPVEMTGKHQLVVRRDYSDSTNPGPPKTPPIAVQKISQ-YACQRRTTLNNC---NQIFTDAFDI-LAENCEFRENEDSCVTFMRAASVLKSLPF 200
DPase beta                       (exon 2) 21 <elanfeknvsqaihkyna-yr> 40                           (exon 3) 41 <AASVIAKYPH  47
repl. MS2        100 DPDVLQTSIRSLIGNLLSGYRSSLFGQCT-FSNGAPMGHKLQDAAPYKKFAE-QATVTPRALRAAL--LVRDQCAAWI--RHAVRY--NESYEFRLVVGNGVFTVPKN 198
ATPase alpha         VGRGLLGRVVNTLGAPIDGKGPLDHDGFSAVEAIAPGVIERQSVDQPVQIGYKAVDSMIPIGRGQR--ELIIGDRQTGKTALAIDAIINQRDSGIKCIYVAIGQKAST
  "    beta          VGKATLGRIMNDLGEPVDMKGEIGEEERWAIHRAAPSYEELSNSQELLETGIKVIDLMCPFAKGGK--VGLFGGAGVGKTVNMMELIRNIAIEHSGYSVFAGVGERTR
..............................................................................................................................
spo2 DPase          RISKFAINNR----------------ISKFEPLESEHKLwALDQEINDRGVR---ID-VDLVKHGIGCDEQYQAGLIAEAKKLT-GLPNPNRTAQLKKWLEEKGLTTS
T7 DPase        156 DGMEWWNFNE-EMMDYNVQDVVV-------------TKAL--LEKLLSDKHYFPPEIDFTDVG-YTTFWSE-SLE--------------AVDIEHRA-AWLLAKQERNG 232
Pol I           370 -GLSFAIEPGVAAYIPVAHDYLDAPDqisreraLELLKPL--LEDEKALKVGQNLKYDRGILANYGIELRGIAFDTMLESYILNS-VAGRHDMDSLAERWLKHKTITFE 475
TdT             201 -TIISMK--DTEGIPCLGSKVKGIIE-----EIIEDGESS-EVKAVLNDERYQSFKLFTSVFGVGL-KTSEKWFR-MGFRTLSKV-RSDKSLKFTRMQKAGFLYYEDLV 300
DPase beta       48 -KIKSGA--EAKKLPGVGTKIAEKID-----EFLATGKLR-KLEKIRQDDTSSSINFLTRVTGIGP-SAARKLVD-EGIKTLEDL-RKNE-DKLNHHQRIGLKYFEDFE 147
repl. MS2       199 NKIDRAACKEPDMNMYLQKGVGAFIR-----RRLKSVGID-LNDQSINQRLAQQGSVDGSLATIDLSSA---------------------SDSISDRLVWSFLPPELYS 280
ATPase alpha        ISNVVRKLEEHGALANTIVVVATASE-----SAALQYLAR-MPVA-LMGEYFRDRGEDALIIYDDLSKQAvay---------------RQ--SLLLRRPPGREAFPGDVFY
  "    beta         EGNDFYHEM-TD--SNVIDKVSLV--------YGQMNEPP-GNRLRVALTGLTMAEKFRDEGRDVLlf-----VDNITRYTLAG---TEVSALLGRMPSA------VGY
..............................................................................................................................
Pol I              476 EIAGKGKNQLTFNQIALEEAGRYAAEDADVT-LQLHL----------------KMWPDL-QKHKGPLNVFENIEMPLV-PVLSRIERNGVKIDPKVLHNHSEELTL-RLAE
spo2 DPase             SLA-KDKIEELIENTNDETVHRVLRL---------------------------RQEM-AK-TSVKKYLAMEK-ALC-PDNRVR-GLLQFYGASRTGRWAGRLVQVQNLP
T7 DPase           233 FPFDTKAIEELYVELAARRSELLRKLTETFC-SWYQP----------------KGGTEM-FCHPRTGKPLPKYPRIKT-PKVGCI-FKKPKNKAQREGREPCELDT-REYV
TdT                301 SCVTRAEAEAVSVLVKEAVWAFLPDAFVTMTGGFRRGK---------------KMGHDVDFLITSPGST-EDEEQ--LLQKVMNLWEKKGLLLYYDL-VESTFEKL-RLPS
DPase beta             KRIPREEMLQMQDIVLNEVKKLDPEYIATVCGSFRRGAE--------------SSG-DMDVLLTHPNFTSESSKQPKLLHRVVEQLQKVRFITDTRSKGETKFMGVCQLPS
repl. MS2          281 YL-dRIRSHYGIVDGETIRWEL----FSTMGNGFTFELESMIFWAIVKATDIHFGNAGTIGIYGDDIiCPSEIAPRVLEALAYYGFK-PNLRKTF--VSGLFRESCGAHF
ATPase alpha           LHSRLLEMLQRVNAEYVE--AFTKGEVKGKTGSLTA-LPIIETQAGDVSA-FVPTNVISITDGQIFLETNLFNAGIRPAVNPGISVSRVGGAAQTKIMKKLSGGIRTALA
  "    beta            QPTLAEEMGVLQ---------ERIT---STKTGSITS-VQAVYVPA-DDLTDPSPATTFAHLDATVVLSRQIASLGIYPAVDPLDSTSRQLDPLVVGQEHYDTARGVQSIL
..............................................................................................................................
Pol I                 -LEKKAHEIAGE--------------------------------------EFN--LSSTKQLQTILFE-KQGIKPLKKTPGGAPSTSEE-VLQELALDYPLPK----VIL 428
spo2 DPase            -QNK-IEDLDTA--------------------------------------R-N--LLKGGHYEAIELL--YGQVPFVLSQLVRTAFIPS-EGNEFYVSDF---SAIEARV
T7 DPase              -AGAPYTPVEHV--------------------------------------VFN--PSSRDHIQKKLQE--AGWVPTKYTDKGAPVVDDE-VLQGVRVDDPEKQAAIDLIK 397
TdT                   -RKVDALDHFQKCFLIFKLPRQRVD-SDQSSWQ--EG-------KTWKAIRVDLVLCPYERRAFAL--LGWTGSRFERDLRRYATHERKMILDNHALYDKTKRIFLKA 486
DPase beta            ENDENEYPHRRI--------------------------------------DIR--LIPKDQYYC--GV--LYFTGSDIFNKNMRAHALE-KGFTINEYTIRPLGVTGVAG 308
repl. MS2             YRGVDVKP-FYIKKPVN-LFALMLILNRLRGWGVVGGMSDPRLYKVWVRLSSQVPSMFFGGTDLAA--DYYVVSPPTASVYTKTPYG-RLLADTRTSGFRLARI--AR
ATPase alpha          Q-YR-ELAAFSQ-------FASDLD-KATRK-QLDHG--Q-KV----TEL---LKQKQYAPMSVAQ--QSL-VLF-AAE-RGYladv-ELSKIGSFEAALLAYV--DR
  "  beta             QRYQ-ELKDIIA--------ILGMDELSEEDK-LVVA---RAR--KI-QRF---LSQPFFVAEVFTG--SPGKYVSLKDTIRGF-KGI--MEGEYDHLPEQAFYM--IE
..............................................................................................................................
Pol I           429   EYRGLAKL-KST-YTDKLPLMINPKTGR--VHTSYHQAVTATG/ 653
spo2 DPase            IAWLADFSAIEARVI-IAWLAGEWRLEVFNTHGKIYEASAAQMFKVPVCSITKGSP-LRQKGKV/ 452
T7 DPase        398   EYLMIQKRIGQSAEGDKAWLRTVAEDGK--IHGSVNPNGAVTGRATHAFPNLAQIPGVRSPYGE/ 449
TdT             487   ESEEEIFAHLGLDYI-----------------------------------EPWERNAQ 508
DPase beta      309   EPLPVDSEQDIFDYI--QW-------------------------------RYREPKDRSEQ 335 (exon 3)
repl. MS2             ERKFFSEKHDSGRYI--AWFH-TGGEITDSMKSAGVRVIRTSEVLTPV-PTFPQECGPASSPRQ 544
ATPase alpha          dHAPLMQE-INQ----------TGGYNDE-IEGKLKG-ILDS-FKATQSWQ 513
  "  beta             -EA---VE-KAK----------KLQ  460
```

Fig. 10. (Next page) Alignment of nucleic acid replicases and F_1-ATPase alpha and beta subunits. Lower case letters are tentatively aligned aa residues. "O" = stop codon; EC = E. coli; DPase = DNA polymerase; Pol I = DNA polymerase I; repl. = RNA replicase (beta subunit); TdT = terminal deoxynucleotidyl transferase; spO2, T7, MS2 = spO2 phage, T7 phage, MS2 RNA phage, respectively. Amino acid matches of spO2 and Pol I to T7 DPase, and of TdT to DPase beta, are underlined. Underlines in other sequences indicate matches to TdT and/or DPase beta. Data are from NBRF Protein Sequence database. Ollis *et al*. (1985) and Ohnishi (1987) are referred.

Class S13	S4 S11 S13 P_{alpha},
	S5 (1) L6 S8 S14 L5 (2) P_{spc},
	S12 P_{STR},
	L10 P_{beta} L1 L11 P_{L11}
	S2
Class S16	S16, S17
Class S7	S7, S19
Class S6	S6, S10
Class S3	S3, S9
Class L23	L23 (2) L22 (2) P_{S10},
	L14, L17
Class L33	L29, L33; L32, L30, L34
Class L7/L12	L7/L12, L24
Class L15	L15, L16
Class L21	L21, L25

Fig. 11. Tentative classification of E. coli robosomal proteins shown together with operon structures. Underlined r-proteins are co-transcribed and share a common promoter denoted by "P". Numerals parenthesized in operon denote the number of r-protein genes existing between the two r-protein genes. This classification is based upon homology relationship among their aa sequences and taken from Ohnishi (1984). L34 is further added to Class L33, since significant homology between L34 and L30 can be found (Ohnishi, in preparation).

gene(s). Thus it can be proposed that a 'small ribosome' consisting of very few (or only one) primitive ancestral r-protein(s) must have existed as a primitive protein-synthesizing aparatus."

This conclusion has now becomes to be more reasonable by the finding (of the primodial gene theory) that EC L34 seems to be the most plausible candidate for the first r-protein, from which all of the other EC r-proteins (probably exepting S1) have evolved by repeated gene duplications. In the later section, origin of the gene (rpmH) encoding EC L34 from an RNA without protein-coding information will be discussed. The first ribosome was most probably a small ribosome consisting of a primitive L34 and (relatively small-sized (?)) rRNA(s).

7. EVOLUTION FROM PROTEIN-NON-ENCODING RO's TO PROTEIN-ENCODING mRNA's

In order to elucidate the evolutionary origin of peptide-encoding gene sequence from peptide-non-encoding sequence, these primodial genes (encoding AK domains and L34 in EC) as well as the rnpA gene coding for the protein component of ribonuclease (RNase) P were aligned with LTSV-A RNA 2 virusoid and the rrnC gene encoding $tRNA^{Trp}$. RNase P seems to be very primitive since its RNA component called "M1 RNA" catalyzes splicing of pre-tRNA with the

```
[ A ]
rrnC, tRNATrp (EC)   5'(1)/AGGGGCGTAGTTCAATT-GGTAGA----GCACCGGTCTCCAAAACCGGGTGTTGGGAGTTCGAGTCTCTCCGCCCCTGCCAgaaa/(80)
 vs LTSV (=) [50.0%]#     == === ==       = ===    =  === ==    =    = =    =    = ==== =======  ==    = ==
LTSV-A RNA 2          (1)/-GAGCUGCGCAG--GGGGC-CG-AGAUUUUGAUUCGGACUGG-UAGGAAGAUAG-GAUGGAUCGACUCUCUCCAGCCUACCAAGUUG/(80)

M1 RNA               5'(1) GAAGCUGACCAG--ACAGU-CGCCGCUUCGUCGUCGUCCUCU-UCGGGGGAGACGGCGGAGGGGAG--GAAAGUCCGGGCUCCAUA/(81)
 vs LTSV (=) [41.8%]#     =====  ===     =  ==  = ==      ===  ==   = ==  == =  =  ===  =          ==    =     =
 vs rrnC (+) [37.7%]      ++  +   ++     + + ++  +      +  ++  +++       ++      ++  +   +        + ++   ++    + +
AK domain II (EC)   73 / I  -   A  Q    E  P* C  R  N  G     F  L   L   D  G* F*  P* R* T  I   P  Q  A  D  A  M / 97
 ( adk gene )          /ATC----GCGCAG--GAAGACTGCCGTAATGGTT---TCCTGT-T-GGACGGCTT-CCCGCGTACCAT-TCCGCAGGCAGACGCGATG/
 vs LTSV (=) [46.5%]#          ======  = =    =     = = =    ===  = === =           =  =     = ==  === =   =   = ==
 vs rrnC (+) [35.8%]           +++ ++   ++    +  +     +       ++          ++  +        +     + ++  +   +   +  ++
 vs M1 RNA(o)[40.3%]#          o  ooo    oo   oooo     o     oooo o o oo  o       oo       o         o o o o oo
RNase P (EC)        27 / R  A  G  T* P  Q* I   T  I  L* G  R* L  N   S* L  G* H  P*  R* I  G   L  T  V  A* K  K* N*/55
 ( rnpA gene )         /CGGGCTGGCACGCCGCAAATTA-CCATTCTCGGCCGCCTGAAT-TCGCTGGGGCATCCC-CGTATCGG-TCTTACAGTCGCCAAGAAAAAC/
 vs L34 (:) [46.4 %]#     :    ::::    :::       : ::    :::     : :: :  ::     ::    :            : :: :  :::::
 vs LTSV (=) [24.7%]     =    == =              = = =              = =   =           =       ===        =====
 vs rrnC (+) [35.5%]        +++      +++ +         +  +   +         + +++    +       +    +  +++       +  + +++++
 vs M1 RNA(o)[27.8%]   o     o  o     o o  oo  o              o ooo  oo o     o         o      o     o o   o o
 vs AK II (x)[41.1%]#        xx x     xx     xx x  x x          x x x   xx     xxx xxxx x    xx      x  x x  xx
L34 (EC)             1   M  K  R  T* F  Q* P*  S  V  L* K  R* N  R   S* H  G* F*  R  A  R  M   -  -  -  A* T  K* N*/26
 ( rpmH gene )         /ATGAAACGCACTTTTCAACCGT-CTGTACTGAAGCGCAACCGT-TCTCACGGCTT-CCGTGCTCGTAT----------GGCTACTAAAAAT/
 vs LTSV (=) [31.4%]     ==   == =       =     =   =  =    == =  =    = =         = ===  =           = =
 vs rrnC (+) [35.3%]      +   ++    +++++      +         +    +        +  ++  +    + + +++  +                 +++
 vs M1 RNA(o)[27.8%]      o    o  o      o      o o    o  o   oo  o oo    o         o  o o          oo    o o o
 vs AK II (x)[50.0%]#  xx      xx x     xx    x x xxx xx       x xx x   xxxxxxx xx    x    xx          x    x x
.....................................................................................................................
[ B ]   ribosomal protein L34 (EC)       1 MKRTFQPSVLKRNRSHGFRARM---ATKNGRQVLARRRAKGrarltvsk 46          aa match

        RNase (barnase) (BAM)       1 AQVINTFSDGVADYLQTY--HKLPNDYITKSEAQALGWVASKGNLADVAPG-ksi/ 51
         vs L34 (+)                        ++   +              ++   + +     ++                     9/44 = 20.5%#
        RNase P (EC)               25 /PQRAGTPQITILGRLNSLGHPRIGLTVAKKNVRRAHERNRIKRLTRESFRLRQHE/ 79
         vs L34 (+)                        + +    + +  + +         + ++ +    + + +    +           14/46 = 30.4%#
         vs barnase                     o   o          o           o     o      o                  6/51 = 11.8%
        RNase H (EC)               29 /RGREKTFSAG---YTRTT--NNRMELMAAIVALEALKEHCEVILSTDSQYVRQGI/ 78
         vs L34 (+)                        ++         +     ++   +      +                          7/41 = 17.1%
         vs barnase (o)                    ooo o   o  o                oo         o       o       10/48 = 20.1%#
         vs RNase P (=)                 =  =                     =     =        =   =   ==         8/49 = 16.3%
```

Fig. 12. Evolution from peptide-non-coding RNA to peptide-coding primodial genes. [A] Alignment of RNAs and primitive DNA genes. Lower case letters in rrnC (EC) indicate residues in spacer. Base matches are shown in the Figure, and percent base matches are given in brackets. "#" denotes ">40%". AK II = domain II of EC adenylate kinase defined by Ohnishi (1986a,b). EC = E. coli. [B] Alignment of RNases with r-protein L34 (EC). Base matches and percent base matches are given in the Figure ("#" denotes ">20%"). Amino acid matches are asterisked. "#" indicates segment of mRNA ribo-organism transcribed from DNA gene encoding ca. 47-aa-long domain of primitive enzymes. Data in [A] and [B] are from: M1 RNA = Reed et al., 1983; Sakamoto et al., 1983; rnpA & rpmH genes = Hansen et al., 1989; barnase = Hartley & Barker, 1972; RNase H = Kanaya & Crouch, 1983; adk gene = Brune et al., 1985.

aid of its 119-aa protein component. The EC M1 RNA was, therefore, also compared with these RNAs and DNAs.

The resulting alignment shown in Fig. 12[A] elucidated an astonishing feature of the evolutionary relationship between peptide-non-coding RNAs and peptide-coding primodial genes. The results elucidate;

(i) Bases 1-81 in LTSV RNA 2 is highly homologous to $tRNA^{Trp}$ (50% base match), suggesting that virusoid also has a multi-ur-RNA (or multi-tRNA ?) structure.
(ii) M1 RNA shows 42% and 38% homology to the virusoid and $tRNA^{Trp}$, respectively, suggesting that M1 RNA seems to be more closely related to the virusoid than to $tRNA^{Trp}$ (as will be further confirmed by Fig. 13).
(iii) An adk gene segment encoding AK domain II shows marked homologies (40-48%) to LTSV-A RNA 2, M1 RNA and rnpA gene (encoding RNase P protein), but shows a homology as weak as 35% to $tRNA^{Trp}$. L34 gene gives 49% base match to adk gene segment.
(iv) As shown in Fig. 12[B], alignment of amino acid sequence of L34 with RNases including barnase (extracellular RNase) (Hartley and Barker, 1972), RNase P protein, and RNase H further confirms that RNase P protein has 30% amino acid match to L34, and is undoubtedly a product of a very primitive gene.

Considering that rpmH gene coding for RNase P protein is mapped at the immediate upstream of rnpA gene coding for L34 r-protein on EC chromosome (Hansen et al., 1985), the close homology between these two genes with M1 RNA seems to be of critical importance in the origin of peptide-coding polynucleotides from peptide-non-coding RNA.

Further comparison of the EC M1 RNA with virusoids (RNA 2) of LTSV-A and SNMV (Solanum nodiflorum mottle virus) revealed that the whole sequence of M1 RNA is indeed homologous to the entire sequences of these virusoids, LTSV-A RNA 2 and SNMV RNA 2 (Fig. 13). The co-ancestor of M1 RNA and virusoids must, therefore, have a highly base-pared rod-like structure as found in contemporary virusoids, because evolution from the nearly completely base-pared structure in virusoids to less complementary structure in M1 RNA could have occurred easily "by breaking of complementary symmetricity", but the evolution in the vice direction would have been very difficult to occur.

We can thus have the following resuts;

(1) mRNAs coding for L34, AK II domain, and RNase P protein domain are close homologues of M1 RNA, virusoids and $tRNA^{Trp}$.
(2) The first primodial mRNA encoding most primitive protein(s) evolved either from an ancestral M1 RNA or from some tRNA-line RO. The most plausible evolutionary tree of ROs is given in Fig. 14-A,
(3) M1 RNA and virusoids share a common ancestral rod-like, highly base-pared RO with an approximate length of contemporary M1-RNA. This common ancestral RO was most probably a primitive M1 RNA with RNase P-like activity, because contemporary virusoids replicate themselves by the use of host RNA polymerase protein made by host's peptide-coding system.
(4) The original rod-like, highly base-pared structure in the ancestral M1 RNA must have "lost its complementary symmetry" by interacting with RNase P protein component throughout evolution to the contemporary M1 RNA (RNA component of RNase P), whereas such breaking of complementary symmetry must have not occurred throughout the evolutionary course from the ancestral M1 RNA to virusoids and viroids. (See Fig. 3 for complementary symmetry.)
(5) This ancestral M1 RNA had a multi-ur-RNA or multi-(proto-)tRNA structure, and was probably very similar to the multi-tRNA/multi-ur-RNA structure of pre-tRNAs, pre-rRNA, and virusoids. The last common ancestor of pre-tRNAs, pre-rRNA and the ancestral M1 RNA was either a primitive multi-ur-RNA RO or multi-(proto-)tRNA RO. A most important problem to be

```
SNMV RNA 2         (1)  GUUCCUGCCCUUGGGGACU--GAUUUUUGGUUCG-CCUGGUCCGUGUCCGUCAGUGGAUGUGUAUCCACUCUGAUGAGUCCAAAGG-ACGAAACGGAUGU- 97
 vs M1 RNA (+)          +  +++ ++     + +  + ++  + + +++ +++  ++ + +         + +++ + +    + ++    +   +++  +++    +     + ++
 vs LTSV-A (:)          :  :::: :  :::: :  ::::::    :::  ::::: :          ::::: :  ::  :::  :      :::  ::     ::    ::  :
LTSV-A RNA 2       (1)  G-AGCUGCGCAGGGGGCCGA-GAUUUUGAU-UCGGACUGGUAGGAAGAUA-GGAUGGAUC-GACUCU-CUC-CAGCCUACCA--AGU-UGGACUUGGUGGGG 92
 vs M1 RNA (+)          + +++++  +++    + ++  + ++      +++  ++  + ++  ++ + ++  +++    ++        ++ + + ++ +++  ++     +   +    ++
M1 RNA (E.coli) (5')1   GAAGCUGACCAGACAGUCGCCGCUUCGUCG-UCGUCCUCUUCGGGGAGACGGCGGAGGGGAGGAAAGUC-CGGGCU-CCAU-AGGCAGGGUGCCAGGUA 98

.........................................................................................................................

SNMV RNA 2          98  ACCGCUUCUUGUCUCGACCU--CGACCUGGACUAGUGAUCGAGGGAGGCUCACCUUCACGCCCGCUGAGU-AGAUGUAGUCGCA-UACUCCAAUGACUUG 193
 vs M1 RNA (+)          ++  ++     +       ++    +++++ +  + +  ++  +        ++  +  +    + + +++  ++    +++    ++       +        ++   +
 vs LTSV-A (:)          :   :   :::     ::: :  :: :::: :                   :    : :    :    :: ::: :  ::    :: :
LTSV-A RNA 2        93  AGACCCAAUUGAGCGGACGUUUCGGCCUGC-CA--------------------UGGCCUCAU-CA--GUGAGACGGCGUUUCAGUCA------------ 155
 vs M1 RNA (+)          +  ++      + +     +     ++ ++ +  ++                    +++++ +    ++   + + ++     ++      ++ +
M1 RNA              99  ACGCCUGGGGGGGAAACCCA--CGACCAGUGCAACAGAGAGCAAACCGCCGA--UGGCCCCGCGCA-AGCGGGAUCAGGUAAGGGUGAAAGGGUGCGGUA 192

.........................................................................................................................

SNMV RNA 2         194  GGGUCACUGUGUAAACGUACUA-CAGAUCUACGGCCAUGUG-ACUGG-CGGGGAGCUGGACCCUCUCACCCCCUCUCAcc--ACUC------AGGUAGUG 272
 vs M1 RNA (+)           +  + +    +        +  ++ +++ ++    +++    ++   +++    +   +++++  + ++     + +       ++  +     ++          ++  ++
 vs LTSV-A (:)                             :: ::   :  :::    ::      :    : :  :    ::     : :::   : :    :   :  : :  ::::          ::  :
LTSV-A   RNA 2     156  ----------------GU-CUCAUACGUCUGA-GCGUGAUA-CCCGCUCACUGAAGAUGGCCCGGU-AGGGC--CGAAACGUACUCAAUUGAGGGGGCUA 233
 vs M1 RNA (+)                             +  + + +  ++ +  ++   + ++   +   + + + ++  + ++++    + ++++    +   ++ ++++          +++    ++
M1 RNA             193  AGAGCGCACCGCGCGGCUGGUAACAG-UCCGUGGCACGGUAAACUCCACCCGGAGCAAGGCCAAAU-AGGGGUUCAUAAGGUAC--------GGGCCGUA 282

.........................................................................................................................

SNMV RNA 2         273  UUGAAGGUCGCGAGGGAG-UCAAGGACGCCCGGCAUCAGAGUGCU-GAUUGCACACCACCGGUAUCACGACUAGGUGGGCUGGCUACCACUCCAGGCUGGCAGGUAAC (378)
 vs M1 RNA (+)           ++++     +  +++  + +  ++     +    +  ++      +  ++ +++       ++     +         ++++++ ++        + ++++       ++  + + +
 vs LTSV-A (:)           :  :     :    :        :  :::  :     :  :  : :     :  :::     :   :    :          ::    : :  :     ::: :  :      ::::::: :
LTSV-A   RNA 2     234  CUUCACCCGGGAUGACCCCUGGAGGUGGAUAGUAACGACAUCCCGAGAUCUGAGCCUUUC-----CA-auC-AAAU--CCUGUCAUCG--UCCAGGCAGCUUC----- (324)
 vs M1 RNA (+)          ++  +++++++  + +  ++ +++    +  ++  +    +    ++ ++++      + ++ ++       ++     + + +     ++ + + +     ++  + +++ +++
M1 RNA             283  CUGAACCCGGGUAGGCUGCUUGAGCCAGUGAGCGAUUGCUGGCCUAGAUGAAUGACUGUC-----CACGAC-AGAA--CCCGGCUUa---uc-GGUCAGUUUCaccu-  377 (3')
```

Fig. 13. Alignment of M1 RNA of *E. coli* ribonuclease P against virusoids. Non-circular M1 RNA (377 bases) is aligned against circular, single-stranded, highly base-paired, rod-like RNAs (RNA 2) (virusoids) of *Solanum nodiflorum* mottle virus (SNMV) and the strain A of lucerne transient streak virus (LTSV-A) (378 and 324 bases, respectively). Sequence data are from Franski (1987) (SNMV RNA 2) and Keese *et al.* (1983) (LTSV-A) and from Sakamoto *et al.* (1983) (M1 RNA). Position 1 in LTSV-A RNA 2 corresponds to position 2 in original numbering of Keese *et al.* (1983). Base matches are indicated in the Figure. Base matches to M1 RNA and to LTSV-A RNA 2 are indicated by (+) and (:), respectively.

solved is whether the fundamental structure of virusoids (and M1 RNA) is the same multi-tRNA structure as found in pre-rRNA and pre-tRNAs (Fig. 14-C), or a multi-ur-RNA structure significantly differing from the multi-tRNA structure (Fig. 14-B). We can not answer to this question at the present status, but the latter case (Fig.14-B) seems to be more reasonable than the former (Fig. 14-C) since some differnces seems to exist between repeating units of pre-rRNA and pre-tRNAs and those of virusoids.

The result (2) allows us to find an important unavoidable condition which "any" hypothesis on the origin of genetic code must satisfy.

As discussed in relation to the "small ribosome theory" (Ohnishi, 1984b) in section 6, ribosome originated most probably as a small ribosome consisting of primitive rRNA(s) and a sole, primitive r-protein directly ancestral to EC L34. Primitive rRNA(s) complexed with L34 must, therefore, have been the first primodial ribosome, which very closely resembles to the contemporary RNase P, since the M1 RNA and rnpA gene are closely homologous to rRNAs and L34 (rpmH) gene, respectively. The first ribosome might have emerged as a kind of RNase P-like ribozyme (or RO) with some activity of interaction with RNA(s) such as ancestral pre-tRNA/pre-rRNA ROs.

Accordingly, the above-mentioned results are well consistent with the folowing hypothesis, [H2], that

[H 2]: genetic informations to code for protein sequences have originated as a cultural product of the ribo-organismic society by mutual and/or altruistic co-operative works of different ribo-organismic species including primitive tRNA ROs, primitive rRNA ROs, and primitive M1-RNA/virusoid RO.

The symbiotic hypothesis of cell machinery ([H 1] in section 4) has thus become further substantiated by this plausibility of [H 2]. [H 1] and [H-2] are, therefore, most reasonable hypotheses to explain for the evolution from primitive ROs to contemporary uni-cellular prokaryotic organisms. Eigen's (1978) theory that tRNA was the first gene has now been proved to be essentially true in somewhat modified fashion.

8. EVOLUTION OF HIERARCHIES IN INDIVIDUALITY AND GENERALIZED CULTURE

8.1. Evolution of Hierarchical levels of individuality

The symbiotic origin of cell machinery concluded above would throw some light upon the problem how the lower level of individuality (e.g. ribo-organismic individuals) could have evolved to generate an upper level of individuality (e.g. cell-individuals). Similar co-operative and/or altruistic symbiotic processes must have caused the evolutionary emergence of every upper level of organic individuality from pre-existing lower level of individuality.

Evolution from uni-cellular haploid organisms to unicellular diploid organisms must have been achieved by (co-operative or some other type) symbiosis of two genetically related unicellular haploid-cell-organisms (Margulis, 1981, p.343; Ohnishi, 1989d).

Furthermore, evolution from uni-cellular diploid organisms to multicellular diploid animals must have been achieved by altruistic symbiosis of of uni-cellular diploid organisms composed of germ-line cell-individuals and somatic-line altruistic cell-individuals (Ohnishi, 1989d). This means that a multi-cellular diploid animal individual is an altruistic society (= a special case of symbiotic society) of unicellular diploid individuals, in which sterile somatic-line cell-individuals altruistically co-operate with fertile germ-line cell-individuals. This altruism must have evolved by the so-called "kinship selection" (Hamilton, 1964a, b). This is because the unicellular diploid cell-organisms constituting an animal body form a kin group

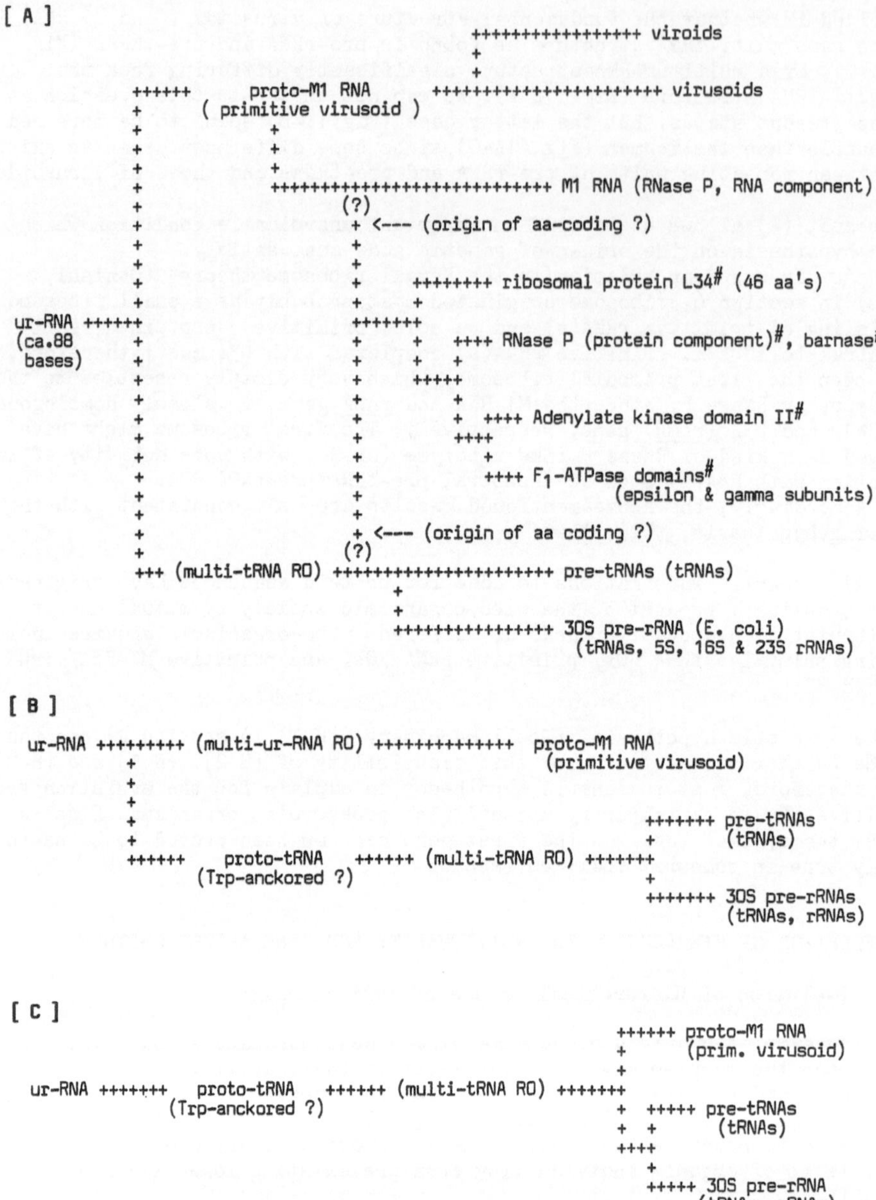

Fig. 14. Tentative evolutionary tree of RNAs including peptide-non-encoding ribo-organisms (RO's) and peptide-encoding primitive mRNA RO's. [A] Evolutionary tree of contemporary RO's. "#" denotes segment of mRNA RO encoding ca. 47-aa-long domain of primitive enzymes. Dual possibility of the origin of peptide-encoding mRNA RO is indicated by "(?)". [B] A most plausible evolutionary relationship among proto-M1 RNA, pre-tRNA, 30S pre-rRNA, and proto-tRNA. In this case, the fundamental structure of M1 RNA and virusoid is multi-ur-RNA, and the fundamental multi-tRNA structure of pre-tRNAs and pre-rRNA have evolved by direct multiplication of proto-tRNA. [C] An alternative, less plausible evolutionary relationship of proto-M1 RNA, pre-tRNA, 30S pre-rRNA, and proto-tRNA. In this case, not only pre-tRNAs and pre-rRNA but also M1 RNA and virusoids have a fundamental structure of multi-proto-tRNA.

(clone) of zygote-derived diploid cell-individuals, and the co-efficient of relationship is unity (r = 1) between every two of these cell-individuals suggesting that highest degree of mutual altruism would be achieved (Hamilton, 1964b, p.25). If we compare this altruistic society (animal body) with the well-known, typical altruistic societies of bees and ants, the germ-line cell-individuals are "queen cell-individuals" corresponding to fertile queen females, and the altruistic somatic-line cell-individuals are "worker cell-individuals" corresponding to sterile worker females, as pointed out by Wilson (1975). This correspondence seems to be completely reasonable, and therefore a common logic must underlie the evolution of these two different hierarchical levels of altruistic society.

Accordingly, in eusocial insects including some hymenopterans and termites, multicellular diploid individuals associate to constitute an upper levelled individual (or society of diploid individuals) called "super-organism" (Wheeler, 1923; Wilson, 1975). It must, however, be noted that, in bees and ants, the logical counterpart of the altruistic diploid-cell society (animal body) consisting of diploid queen-cells and worker-cells is the hymenopteran altruistic female society consisting of diploid queens and workers, and not the society of these females and haploid males. Another type of "super-organism" is a co-operative society or "colony" consisting of kin (clonal) zooids as typically found in Physalia and other colonial animals. Since r = 1 between any two of these zooids, the both types of altruistic or co-operative society of multi-cellular diploid individuals are considered to have evolved, most probably, by kinship selection (Hamilton, 1964b).

Now we find a hierarchy of different levels of individuality as shown in Table 3, in which the upper-level individual evolved as a co-operative or altruistic society of the lower-level individuals. The lower-level individuals are, therefore, members of the society. Most of these co-operative or altruistic societies must have evolved by kinship selection as disscussed above. The cell or cell-machinery is also a co-operative symbiotic society of ribo-organismic individuals. Whether this step of co-operative evolution also occurred by something like kinship-selection of ROs is an important problem remaining to be solved.

8.2. Evolution of hierarchical levels of generalized culture and cultural products

Human society has its own cultures and cultural products, which are the so-called "extended phenotypes" (Dawkins, 1982) more or less satisfying Dawkin's (1976, 1982) definition of "meme" which is a "replicator". Watch-making culture and the resulting watch (cultural product) are memes in human society. Watch is a human-made machine, meaning that it is made by (some) members of human society. Organization of human society is also one kind of culture (or cultural product), and therefore a meme. The human society itself is a product of this culture, and therefore is also another meme. This organization of human society can also be considered as a machinery made by members of the society, and human society itself is a human-made "society-machine". Memes evolve by providing the society members with higher Darwinian fitness to cause advantages in their indirect replication using the society-machinery.

Similarly, multicellular organization of animal body is a "culture" or "generalized culture" of constituent diploid cell-organisms, and the resulting animal body itself is a "society-machine" made by the constituent germ-line and somatic-line cell-individuals. This "animal-body machine" closely resembles to "human-made machine" (e.g. watch) evolutionarily produced by co-operative works of human individuals (also based partly on the co-operative works of nerve cell-organisms). These considerations allow us to analyze what is the most essential feature of human culture comparable to "generalized cultures" of other co-operative societies of various organisms at different hierarchical levels.

Table 3. Tentative comparison of cultures and cultural products of societies at different hierarchical levels of individuality.

hierarchical level of individuality (society member)	cooperative society (society-machine)	culture of the society (or society's function) [cultural product]++ (cooperative work by society's members [machine or tool])
ribo-organism (RNA)	prokaryotic or haploid cell (symbiosis)	protein synthesis [protein] DNA replication [DNA] cell division** [cell]+ mitosis** [haploid cell]+
uni-cellular haploid organism (haploid cell)	diploid cell (2-member-society)# (symbiosis ?)	DNA exchange by crossing-over [diploid cell with new genotype]+ kinship-recognition by chromosome-pairing in meiosis [diploid cell as a co-operative complex of two kin haploid-cell-individuals, and species as an inter-breeding polulation of diploid-organisms] #, + meiosis and conjugation (fertilization)** [diploid cell]+
uni-cellular diploid organism (diploid cell)	animal body *,## (altruistic symbiosis)	morphogenesis [animal forms] embryogenesis** [aninal body]+
multi-cellular animal individual (animal body)	animal society $	animal social behaviours [animal tools, nests, etc.]
human individual (human body)	human society $$	watch-making [watch] ## computer-making [computer] democracy [democratic society]+

(_) Important points discussed here are underlined.
++ Cultural products (which are machines or tools) in the corresponding co-operative society are bracketted. Both culture and cultural products can be considered as to be "memes" of the society.
+ Society-machine.
* Based on altruistic behaviours of sterile somatic diploid cell-organisms to fertile germ-line diploid cell-organisms.
** Ontogenesis of the society (or "sociogenesis").
# Based on the hypothesis presented in these Proceedings (Ohnishi, 1989d).
See Dawkin's "Blind Watchmaker" (Dawkins, 1986).
$ Upper levelled individuality is acquired in some cases of the societies of eusocial insects (bees, ants, aphids and termites).
$$ Upper-levelled individuality is not acquired.

Table 4. Machines (or tools) and machine-makers

watch:	human-made machine *
human society:	human-made society-machine
animal body:	diploid-cell-organism-made society-machine
protein:	ribo-organism-made machine (or tool)
cell (prokaryotic or haploid):	ribo-organism-made society-machine
diploid cell:	haploid-cell-organism-made society-machine

* "X-made machine" means that the machine is made by co-operative work(s) of X'es who are members of the corresponding co-operative society.

As discussed in section 4, prokaryotic or haploid cell is a society-machine made by ROs including tRNA ROs, rRNA ROs, M1 RNA RO, and mRNA ROs. Protein-synthesizing mechanism is an example of the generalized culture in this mutually co-operative ribo-organismic society. Proteins are "ribo-organism-made machines or tools", whereas conventional machines (such as watch and computer) in our human society are "human-made machines". Both machines adaptively evolved as "memes" of the respective society by natural and/or artifitial selection, and artifitial selection is a special case of natural selection. The common underlying logic in making both of these machines would be a "true watchmaker" (true maker of "the machine") (Dawkins, 1986) in the co-operative society at the corresponding hierarchical level. DNA gene is a meme created by ribo-organismic cultural society, because DNA gene is a typical replicator machine (or tool) made by ROs. This meme is therefore more or less selfish.

These tentative connsiderations on cultures and cultural products in various hierarchical levels of co-operative/altruistic society are summarized in Table 3. Machines and machine-makers are given in Table 4.

8.3. Origin of diploid species as viewed from kinship-recognizing behaviour of haploid-cell individuals

In the kinship-recognition theory of chromosome-pairing in meiosis (Ohnishi, 1989d), it is postulated that chromosome-pairing in meiosis of diploid organisms must be a kinship-recognizing behaviour of the two haploid-cell individuals constituting the symbiotic uni-cellular diploid-cell or the germ-line diploid cell (which is a two-member society of uni-cellular haploid organisms). Under this hypothesis, diploid species as "groups of inter-breeding natural population" defined by Mayer (1969, p.26) would have reasonably evolved from this type of kinship-recognizing interactions between the haploid-cell individuals in the life cycle of this symbiotic diploid organism having diploid and haploid phases of alternative generations.

Since a diploid cell is a two-member society of haploid-cell organisms, the chromosome-pairing in meiosis must be a culture in this society, by which kinship-recognition of the two constituting haploid-individuals can be efficiently achieved.

9. ORTHO-INDIVIDUALITY AND PARA-INDIVIDUALITY

9.1. Ortho-individuality

In the previous section, it is concluded that hierarchy of individuality is given by

[RO ——> haploid-cell oganism ——> diploid-cell organism ——> multi-cellular animal ——> animal (or human) society], [Hi-1]

where every upper-level individual most probably evolved as a co-operative or altruistic society of the lower-level individuals, as summarized in Table 5. It must be noted that there are neither tissue-level nor organ-level organism in this hierarchy, because there never lived any type of tissue-organisms and organ-organisms as free-living individuals. Tissue and organ are mere structural levels of multi-cellular organisms, and not the levels of individuality of organisms. A tissue-like organism such as a slime mold is not tissue, but a "multi-cellular body" of the organism, because any "multi-tissue organism" did not evolved by co-operative association of free-living one-tissue-organisms, but evolved by modifying its morphogenetic and/or ontogenetic pattern. A scheme of hierarchy of organism such as that of Koestler (1978) given by

[gene ——> cell ——> tissue ——> organ ——> organ system ——> individual ——> ......], [Hi-2]

is a false hierarchy, or of no use.

Table 5. Hierarchical structures in ortho-individuality and para-individuality.

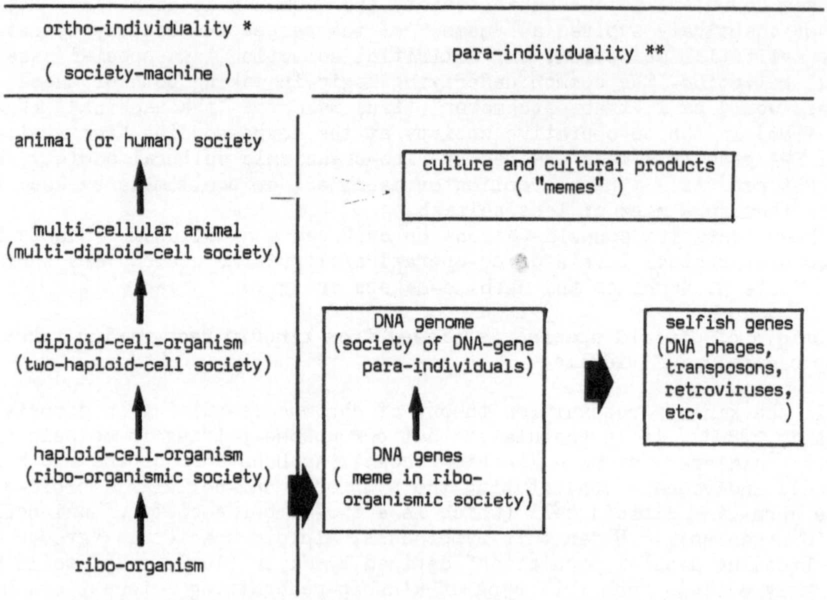

* Ortho-individuals exist or existed as free (or independently) living organisms.
** Para-individuals live depending on ortho-individuals, and have never lived independent of the host ortho-individuals.

```
tRNATrp (EC) :       5'(1)  -AGGGGCGØAGUUCAAØØGGØAGA----GCACCGG-U¢U-CCAaa/(38)
 vs tRNATrp (39-76)[+]          +++ ++ +   + ++ + ++        + +++   ++ +++
tRNATrp (EC) :         (39) /accGGGU¢UUG-GGAGØØCG-AGUC---UCUCCGCCCCUGCCA 3'(76)

LTSV-A RNA 2 :         (1) /GAGCUGCGCAG--GGGGCCG-AGAUUUUGAUUCGG-ACUGGU/ (38)
 vs tRNATrp ( 1-38)[=]       ==  === ==         = ===    =   ===  ==
 vs tRNATrp (39-76)[=]           = =  =  = =  == ==        = ==   ===
 vs LTSV-A  (39-75)[+]       ++        ++   +++ ++ +       + + +  ++
LTSV-A RNA 2 :        (39) /-AGGAAGAUAGGAUGGGUCG-ACUC---UCUCCAG-CCUACCA/  (75)
 vs tRNATrp ( 1-38)[=]       ===    ===       = = =        = == =  == ===
 vs tRNATrp (39-76)[=]         =  = = =     = === = ==   =====   === ===
```

Fig. 15. Alignment of half-sized tRNA domains of $tRNA^{Trp}$ with each other and with LTSV-A RNA2 virusoid sequence segments. Modified based in $tRNA^{Trp}$ are over-printed by slash (/). Lower case letters indicate tentatively aligned residues. Base matches are indicated by + or =. The 3'-terminal triplet "74 CCA 76" or $tRNA^{Trp}$ and its possibly homologous triplets including anti-codon (34 CCA 36) of $tRNA^{Trp}$ are written in bold type. See legend of Fig. 1 for abbreviations and references of original sequence data.

The individuality in [Hi-1] is denoted hereafter by "ortho-individuality" as shown in Table 5.

9.2. Para-individuality and selfish genes

DNA genes are individuals which depend on cell-individual (which is a ribo-organismic society). DNA genome has a hierarchical structure in individuality depending upon the cell-individual, as given by

[gene ——> cistron ——> operon ——> chromosome ——> haploid genome ——> diploid geneome] [Hi-3]

These individuals have never lived independent of ortho-individual (cell-individual or society of ROs), and the individuality of this kind is therefore denoted by "para-individuality". [Hi-3] is hierarchical levels of para-individuality. DNA genes are memes (machines or tools) in ribo-organismic society. Proteins are also ribo-organism-made machines or memes, and therefore are para-individuals. From a more general viewpoint, most memes might well be said to be para-individuals (Table 5.).

On the other hand, viruses are much like individual organisms although their existence depends on other individuals, and seem to have never lived as free-living organisms. Transposases, reverse transcriptases of retroviruses, and DNA polymerases of T7, sp02 and T4 phages are homologous to host DNA polymerase(s) (Pol I of EC) (Ohnishi, 1988b,1989e). It therefore is most plausible to conclude that transposons, retroviruses and some DNA phages must have emerged first as a selfish extra DNA polymerase gene generated by duplication of host DNA polymerase gene, and thereafter evolved further selfishly by increasing their own (virus) genome-size through its gene duplications. Most viruses would probably have more or less similar selfish origin(s).

Host cell DNA polymerase gene has a key role to replicate the whole genome of the host, and therefore seems to be difficult to exist in two or more copies. Accordingly, duplication of host DNA polymerase gene would have inevitably resulted in a fierce competition between the original and newly duplicated genes, resulting in a selfish evolution or extinction of the defeated. This would give a most plausible answer to the unsolved question why the so-called "selfish genes" (Dawkins, 1976, 1982; Doolittle & Sapienza, 1980; Orgel & Crick, 1980) including transposons and retroviruses (and other viruses) are selfish (Ohnishi, 1988b, 1989e).

Bacterial genome is a symbiotic society of DNA-gene para-individuals, and selfish transposons and phages live depending on the cell, a ribo-organismic society having a DNA genome machine. These selfish genes are therefore para-individuals whose life depend on a ortho-individual.

The relationship among these para-individuals and ortho-individuals are schematized in Table 5.

10. ORIGIN OF ca. 88-BASE UR-RNA FROM A HALF-SIZED RNA ORGANISM

The next important problem to be solved is the evolutionary origin of the 88-base-long ur-RNA, which we tentatively call in this section "unit-sized ur-RNA" for convenience, and tentatively denote it by "ur-RNA_{1u}". $tRNA^{Trp}$ is most suitable for the analysis of tRNA origin, because its sequence seems to be more primitive than any other tRNA as mentioned before.

From Fig. 15, we can find that a $tRNA^{Trp}$ molecule consists of two half-sized RNA domains highly mutually homologous to each other, giving 61.8 % base match and $P_{nuc}(21, 34) = .62 \times 10^{-5}$. Thus it is most plausible that a "half-sized ur-RNA" (tentatively denoted here by "ur-$RNA_{u/2}$") with a length of ca. 38 bases existed before the emergence of ur-RNA_{1u}. Comparison was made between these halves of $tRNA^{Trp}$ and the halves of the tRNA-like or ur-RNA-like unit sequence in LTSV-A RNA 2 (positions 1-75) (Fig. 15). Base match between these two halves of this viroid sequence unit is 44% (15/34) and P_{nuc} = .012, which is very weak homology as compared with that found between the two halves of $tRNA^{Trp}$. This finding suggests that $tRNA^{Trp}$ might be more primitive than virusoids.

A more important point found in Fig. 15 is that the 3'-terminal triplet segment "74 CCA 76" in the 3'-half-sized unit of $tRNA^{Trp}$ is conserved as an anticodon triplet (34-36) in the 5'-half-sized unit of $tRNA^{Trp}$, and as "73-75" of the virusoid. This relationship suggests the following conclusions, some of which would have important implications for future study on the origin of triplet codons.

(1) Evolution from proto-tRNA to contemporary $tRNA_{Trp}$ and other tRNAs was a process of symmetry-breaking in Watson-Crick type base-pairing complementarity of RNA sequence.
(2) Thus obtained $ur\text{-}RNA_{u/2}$ does not satisfy the hairpin-like completely base-paired secondary structure (complete complementary symmetry) in Fig. 3-Ia. It is therefore plausible that $ur\text{-}RNA_{u/2}$ did not replicate by the complementary symmetry model (in Fig. 3-II). It more plausibly replicated either by the rolling-circle model or by MS2-type plus-minus method using its own or some other ribozyme activity.
(3) Proto-tRNA had "CCA" at two positions corresponding to the anti-codon region and 3'-terminus of contemporary tRNAs.
(4) The "CCA" in the anti-codon region of proto-tRNA was the origin of the anti-codon "5' CCA 3'" of contemporary $tRNA^{Trp}$.
(5) Proto-tRNA most probably interacted with Trp or Trp-like amino acid (such as some other aromatic or hydrophobic amino acid) by both or either of its 3'-terminal "CCA" and anti-codon region "CCA", but this interaction did not work as peptide-coding information.
(6) $Ur\text{-}RNA_{1u}$ was either the same RO as this proto-tRNA or direct ancestor of it.
(7) $Ur\text{-}RNA_{1u}$ evolved by $ur\text{-}RNA_{u/2}$ having "CCA" in its 3'-terminus.
(8) $Ur\text{-}RNA_{u/2}$ also possibly interacted with Trp or Trp-like amino acid(s) by its 3'-terminal "CCA".
(9) It is also possible that the $ur\text{-}RNA_{u/2}$ RO lived as a dimer of two $ur\text{-}RNA_{u/2}$ individuals. This dimer was probably very similar to $ur\text{-}RNA_{1u}$.

The conclusion (4) is the first discovery of the origin of triplet codon based on substantial evidences.

ACKNOWLEDGEMENTS: The author acknowledges Drs. H. Yanagawa and A. Lazcano for their useful discussions on RNA world.

REFERENCES

Bloch, D., McArthur, B., Widdowson, R., Spector, D., Guimaraes, R.C. and Smith, J., 1983. J. Mol. Evol., 19: 420-428. (See also 1984. Origins Life, 14: 571-578, by the same authors.)
Bloch, D. and Staves, M., 1986. Origins Life, 16: 309-310.
Branch, A.D., Benefeld, B.J. and Robertson, H.D., 1988. Proc. Nat. Acad. Sci. USA, 85: 9128-9132.
Branch, A.D. and Robertson, H.D., 1984. Science, 223: 450-455.
Branch, A.D., Robertson, H.D., and Dickson, E., 1981. Proc. Nat. Acad. Sci. USA, 78: 6381-6385.
Brosius, J., Dull, T.J. and Sleeter, D.D., 1981. J. Mol. Biol., 148:107-127.
Brune, M., Schumann, R. and Wittinghofer, F., 1985. Nucleic Acid Res., 13: 7139-7151.
Cech, T.R., 1989a. Nature (Lond.), 336: 507-508.
Cech, T.R. (ed.), 1989b. "Molecular Biology of RNA", A. R. Liss, New York.
Chen, R., 1976. Hoppe-Seyler's Z. Physiol. Chem., 357: 873-886.
Diener, T.O. (ed.), 1987a. "The Viroids", Plenum, New York.
Diener, 1987b. In: Diener (ed., 1987a), pp.1-5.

Diener, 1987c. In: Diener (ed., 1987a)., pp.9-35.
Dawkins, R., 1976. "The Selfish Gene", Oxford University Press, Oxford.
Dawkins, R., 1982. "The Extended Phenotype", Oxford University Press, Oxford.
Dawkins, R., 1986. "The Blind Watchmaker", Longman, Harlow, Essex.
De Jong, W.W., Gleaves, J.T. and Boulter, D., 1977. J. Mol. Evol., 10 :123 -135.
De Jong, W.W., Zweers, A., Joysey, K.A., Gleaves, J.T. and Boulter, D., 1985. In: "The Evolution and Ecology of Armadillos, Sloths, and Vermilinguas", ed. by G. G. Montgomery, pp. 65-76, Smithsonian Institution Press, Washington and London.
Doolittle, W.F. and Sapienza,C., 1980. Nature (Lond.), 284: 601-603.
Eigen M., 1978. Naturwissenschften, 65: 341-369.
Erdmann, V. and Wolters, J., 1986. Nucleic Acids Res., 14(Suppl.): r1-r59.
Evrard, J.-J., et al., 1988. Gene, 71: 115-122.
Forster, A.C. and Symons, R.H., 1987a. Cell, 49: 211-220.
Forster, A.C. and Symons, R.H., 1987b. Cell, 50: 9-16.
Forster, A.C., Jeffries, A.C., Shelton, C.C. and Symons, R.H., 1987. Cold Spring Harbor Symp. Quant. Biol., 52: 249-259.
Francki, R.I.B., In: Diener (ed., 1987a), pp. 205-218.
Hamilton, W.D., 1964a. J. Theoret. Biol., 7: 1-16.
Hamilton, W.D., 1964b. J. Theoret. Biol., 7: 17-52.
Hartley, R.W. and Barker, E.A., 1972. Nature, New Biol., 235: 15-16.
Hansen, F.G., Hansen, E.B. and Atlung, T., 1985. Gene 38: 85-93.
Haseloff, J. and Symons, R.H., 1981. Nucleic Acids Res., 9: 2741-2752.
Hennig, W., 1966. "Phylogenetic Systematics", University of Illinois Press [Transl. by D. Dwight and R. Zangerl from: Hennig's "Grundzuege einer Theorie der phylogenetischen Systematik" (1950)]
Hennig, W., 1965. Ann. Rev. Entomol., 10: 97-116.
Hirsch, D., 1971. J. Mol. Biol., 58: 439-458.
Joice, G.F., 1989. In: Cech (ed., 1989b), pp. 361-371.
Kanaya,S. and Crouch, R.J., 1983. J. Biol. Chem., 258: 1276-1281.
Keese, P. and Symons, R.H., 1987. In: Diener (ed., 1987a), pp. 37-62.
Keese, P., Bruening, G. and Symons, R.H., 1983. FEBS Letter, 159: 185-190.
Koestler, A., 1978. "Janus ____ A summing up", Huchinson, London.
Lacey, J.C., Jr., Staves, M.P., and Bloch, D.P., 1986. Origins Life, 16: 525 -526.
Margulis, L., 1981. "Symbiosis and Cell Evolution", McGraw-Hill, New York.
Marotta, C.A., Varricchio, F., Smith, I. and Weissman, S.M., 1976. J. Biol. Chem., 251: 3122-3127.
Matsukage, A. et al., 1987., J. Biol. Chem., 262: 8960-8962.
Mayer, E., 1969. "Principles of Systematic Zoology", McGraw-Hill, New York.
Moore, G.W. and Goodman, M, 1977. J. Mol. Evol., 9: 121-130.
Murgora, E.J., Goeringer, H.U., Dahlberg, A.E. and Hijazi, K.A., 1989. In: Cech (ed.), pp. 221-229.
Nazarea, A.D., Bloch, D.P. and Semrau, A.C., 1985. Proc. Nat. Acad. Sci. USA, 82: 5337-5341.
Nairn, C.J. and Ferl, R.J., 1988. J. Mol. Evol., 27: 133-141.
Orgel, L.E. and Crick, F.H.C., 1980. Nature (Lond.), 284:604-607.
Ohnishi, K., 1984a. Origins Life, 14: 707-715.
Ohnishi, K., 1984b. Origins Life, 14: 716-724.
Ohnishi, K., 1986a. Origins Life, 16: 247-248.
Ohnishi, K., 1986b. Nucleic Acids Res., Symp. Ser., No.17, 127-130.
Ohnishi, K., 1986c. Origins Life, 16: 326-227.
Ohnishi, K., 1987. Viva Origino, 16: 69-85.
Ohnishi, K., 1988. Viva Origino, 16: 230-232.
Ohnishi, K., 1988b. Nucleic Acids Res., Symp. Ser., No.20, 97-98.
Ohnishi, K., 1989a. Origins Life, 19(3/4): in press. [two pages with Figures and a Table].
Ohnishi, K., 1989b. In: "Endocytobiology IV. Proceedings of the 4th Inter-

national Colloquium on Endocytobiology and Symbiosis", in press.
Ohnishi, K., 1989c. Origins Life, 19(3/4): in press.[two pages with a Figure of aa sequence alignment].
Ohnishi, K., 1989d. in preparation.
Ohnishi, K., 1989e. Origins Life, 19(3/4): in press. [two pages with a Figure of aa sequence alignment].
Ollis, D.L., Brick, P., Hamlin, R., Xuong, N.G. and Steiz, T.A., 1985. Nature (Lond.), 313: 818-819.
Orgel, L.E., 1987. Cold Spring Harbor Symp. Quant. Biol., 52: 9-16.
Orgel, L.E. and Crick, F.H.C., 1980. Nature (Lond.), 284: 604-607.
Osorio-Almeida, M., Guillemaut, P., Keith, G., Canaday, J. and Well, J.H., 1980. Biochem. Biophys. Res. Comm., 92: 102-108.
Reed, R.E., Bauer, , M.F., Guerrier-Takada, C., Donis-Keller, H. and Altman, S., 1982. Cell 30: 627-636.
Sakamoto, H., Kimura, N. and Shimura, Y., 1983. Proc. Nat. Acad. Sci. USA, 80: 6187-6191, 1983
Schwartz, A.Z., Visscher, J., van der Woerd, R. and Baker, C.G., 1987. Cold Spring Harbor Symp. Quant. Biol., 52: 37-51.
Shiraishi, H. and Shimura, Y., 1988. EMBO J., 7: 3817-3821.
Spinzl, M., Moll, J., Meissner, F. and Hartman, T., 1985. Nucleic Acids Res., 13(Suppl.): r1-r49.
Vogel, H., 1978. J. Mol. Evol., 10: 339-348.
Watson, J.D., Hopkins, N.H., Roberts, J.W., Steitz, J.A. and Weiner, A.M., 1987. "Molecular Biology of the Gene. 4th ed.", Benjamin/Cummings, Menlo Park, California.
Wheeler, W.M., 1923. "Social Life among the Insects", Harcourt, Brace & World, New York.
Wilson, E.O., 1975. "Sociobiology: The New Synthesis", Belknap Press of Harvard University Press, Cambridge, Massachusetts.
Yaniv, M. and Barell, B.G., 1971. Nature (Lond.), 233: 113-114.
Yarus, M. and Barell, B.G., 1971. Biochem. Biophys. Res. Comm., 43: 729-734.
Yong, R.A., 1979. J. Biol. Chem., 254: 12725-12731.

INFLUENCE OF NUCLEIC ACID ON RACEMISATION OF PEPTIDE SYNTHESIS BY WATER SOLUBLE CARBODIIMIDE AND ITS RELEVANCE TO THE ORIGIN OF GENETIC CODE

S.K. Podder[1] and H. S. Basu[2]

Department of Biochemistry
Indian Institute of Science
Bangalore - 560012/India

INTRODUCTION

To understand the principle of complementarity in protein nucleic acid interaction as well as its manifestations in the organization of nucleoprotein complexes and the control of gene regulation and expression is a key problem in the present day of molecular biology(1). It is probably important in understanding chemical evolution of .life based on coupling between proteins and nucleic acids and as well as near universal condon-amino acid relationship. The template directed polymerization of nucleic acids and of amino acids is the key element in contemporary living systems. The rules that govern self recognition of nucleic acid are simple but uniquely defined by the rules of complementarity in the interaction of purines and pyrimidines for base pairing. It is also manifested in template directed in enzymatic (ie replication and transcription) and nonenzymatic polymerization of nucleic acids(2-4). Enzymes are involved only in kinetic control. In contrast nucleic acid directed condensation of amino acids takes place via t-RNA and involves condon-anticodon specificity; the error free incorporation of L-amino acids is rather kinectically controlled(5). Thus it does not give us any clue as to how the intricate machinery of protein synthesis evolved from prebiotic soup containing amino acids/dipeptides, and mononucliotides, whose synthesis are shown to be feasible under prebiotic conditions.

Several suggestions have been made from time to time(6-8). We, however believe that nucleic acid replication and genetic coding of amino acids and peptides are coevolved. It is difficult to address the questions with experimental support without any basic information about the energetic of individual amino acid-base interaction(9-10). Studies on the interaction of regulatory proteins likes repressor with well defined sequences of nucleic acids do suggest that the side chain of amino acids interacts specifically with base pair in major groove, the energetic contribution of the individual base pair-amino acids to overall sequence

1. Laboratory of Mathematical Biology, NCI-FCRF, Frederick, Maryland 21701
2. Brain Tumor Research Center, University of California, San Francisco, California 94143-0520

Symmetries in Science IV
Edited by B. Gruber and J. H. Yopp
Plenum Press, New York, 1990

specific interaction is yet to be estimated(11-12). The higher specificity of interaction is largely electrostatic in nature and is context dependent. Thus we became interested in the study of interaction of the amino acid/dipeptide with mono-, di- and poly- nucleotides with increasing structural and sequence complexity by physio-chemical methods and examined the effect of nucleic acids on carbodiimide mediated condensation of dipeptides. Our objective was two fold: (i) to examine whether the extent of polymerization of a particular amino acid/dipeptide can be selectively enhanced by virtue of its stronger interaction with a specific nucleotide from a mixture of several amino acids/dipeptides and (ii) to assess the changes in the degree of racemisation that occurs in the presence of nucleic acid, is due to the specificity of interaction.

It was long been known that the rate of polymerization of individual units mediated by specific condensing agents is enhanced in the presence of an inert polymer chain. This phenomenon, commonly known as the "Chain Effect", was found to be quite pronounced during the formation of dipeptides from amino acid anhydrides or amino acid dicyanamide mixtures in the presence of polyethylene glycol and different polyamides, like polyglycine, polysarcosine and poly-L-Leucine. To explain this phenomenon of the "Chain Effect" a kinetic model was proposed based on the increase of the local concentration of monomers and the specific orientation effect favorable for condensation induced by the non-covalent association of monomers with the polymer templates(fig.1). The observation that the rate of the condensation reaction in the presence of polyamides was further enhanced by reducing the dielectric constant of the solvent with addition of dioxane indicated that the non-electrostatic interactions, like the hydrogen bonding of amino acids with the NH group of polyamide, play the major role to decide the degree of the enhancement(13-16). Therefore, polymers like, polynucleotides, which are capable of involving in non-covalent interactions with amino acids should, in principle, enhance the yield of amino acid condensation products.

Thus it was necessary for us to collect data for the affinity constant of various amino acids/dipeptides towards mono-, di- and poly-nucleotides. From the measurement of solubility of mononucleotide in solution of various amino acids it was shown that affinity of amino acids toward nucleosides is weak but specific(17). In recent years the interaction of methyl esters of amino acids with mono- and poly-nucleotide has also been studied by NMR technique(18-19). The affinity parameters are found to be in the range.(1-20M^{-1}) In our Laboratory we have undertaken a systematic study of the interaction of dipeptides of the type glyx (x=any other amino acid) with mono-, di- and poly-nucleotide mainly because their zwitter ionic nature would minimize the electrostatic contribution to the overall binding energy and therefore the participation of amino acid side chain in the formation of specific complexes can be demonstrated. Moreover, the glycine is known to be present more abundantly than any other amino acids in primitive atmosphere, the choices of dipeptide appears more relevant. In the sections to follow we shall report the results of these studies to show the specific interaction of dipeptide with mono-, di- and polynucleotide could lead to the selective enhancement of oligopeptide formation in nucleic acid directed polymerization reaction.

Affinity of dipeptides towards nucleic acids

Earlier we reported the affinity parameter (KM^{-1}) of 1:1 complexes of various dipeptide-mononucleotide systems, determined by NMR technique in the range (6-30M^{-1}); the highest value is found for 1:1 CMP: gly-gly(20). Similar studies with different dinucleotide monophosphate with various dipeptides are yet to be made. The complex of GpC with gly-phe and gly-tyr has been reported to have a 10-20 fold higher value as

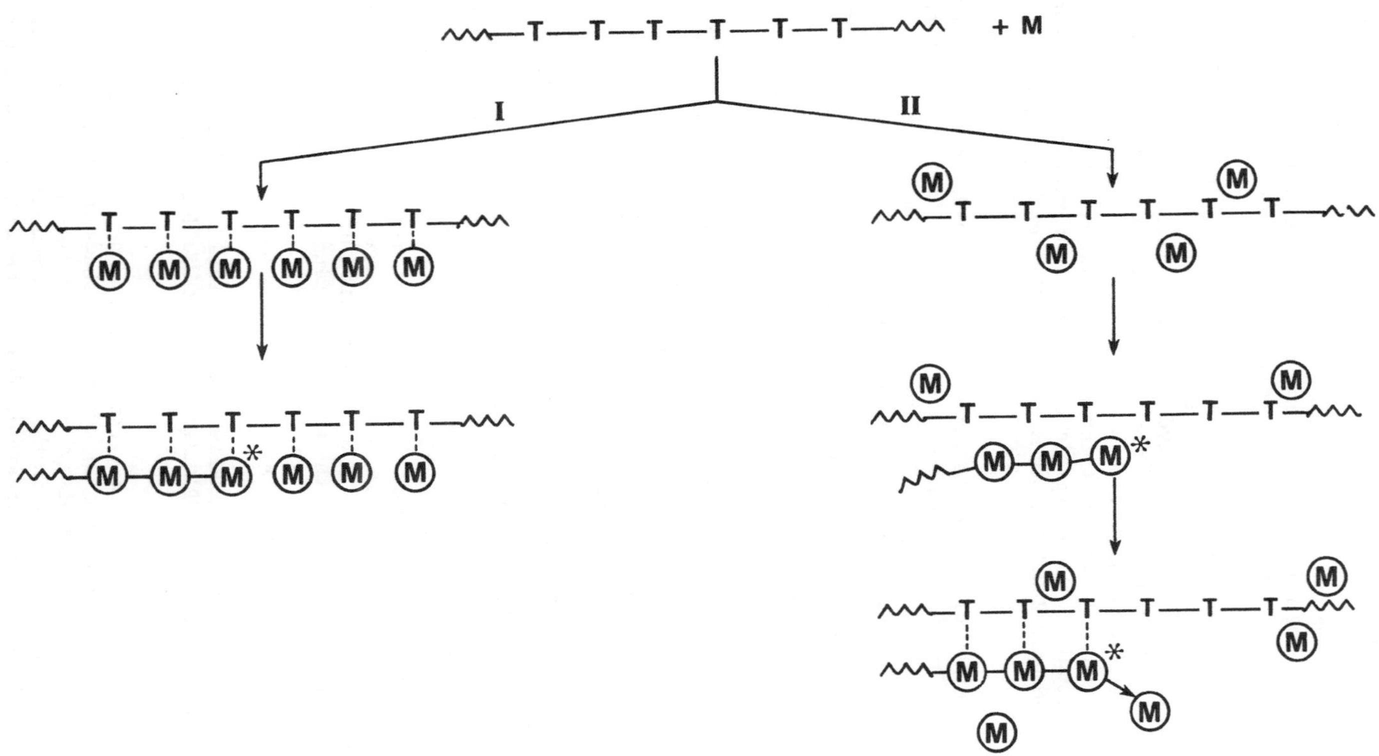

Fig. 1. Model for rate enhancement in template dependent reactions (M) and ∿∿ -T-T- ∿∿ represent monomer and template.

excepted(21). In general the interaction of D amino acid is weak, as can be seen from the table-1 and thus affinity column chromatography was used throughout this study for estimating the extent of racemisation in condensation reaction. (KM^{-1}) values are found to be dependent on (i) nature of bases and amino acid side chains(ii) sugar residue (iii) location of phosphate group and (iv) chirality. Evidently the factors responsible for giving rise to such selectivity must lie in the site binding model involving noncovalent interactions such as H-bond formation between functional group of amino acids residues and bases, and hydrophobic and stacking interactions between them, in addition to electrostatic interaction between the charged groups; The participation of exocyclic amino group of nucleotide was evident from the observed changes in the kinetics of formaldehyde adduct formation with mono- and di-nucleotides in the presence of dipeptides. Assuming the protonated amino group interacts strongly with negatively charged phosphate moiety of nucleotides, various nonbonded interactions can be identified from model building studies; the energetic contribution of various nonbonded interaction is about,0.2-0.5 kcal/mole where as that of NH_3^+-PO_4 is about 1kcal/mole(22).

The interaction of dipeptides with double stranded DNA was studied by measuring the Tm of DNA as shown in figure 2. One should expect an increase in Tm due to charge neutralization as (NH_2) groups are protonated at pH-7. For some dipeptides a decrease in Tm was observed. Such behavior was also observed with amides of amino acids(23). Hence data were analyzed according to following site binding model(24).

$$HL \gtrless L+H \gtrless C+L \gtrless CL$$

The ligand dipeptide (L) may bind to helix H and coil C. According to this model the ligand concentration dependence of Tm is given by equation 1.

$$\frac{1}{T_m^{\circ}} - \frac{1}{T_m} = \frac{R}{\Delta H} \ln \frac{(1 + K_H \cdot L)^{2/n_h}}{(1 + K_c \cdot L)^{2/n_c}} \qquad (1)$$

When T_m° is the melting temperature in the absence of ligand, ΔH enthalpy change due to melting of a base pair, K and n's are binding constant of ligand and number of bases per binding site,(the subscript, H and C represent nucleic acid as helical and coiled from), R gas constant. The values of binding constant of dipeptide and amides of amino acids to poly nucleotide are also given in the table-1.

It is seen that these values are higher than reported earlier for dipeptide-mononucleotide complexes because of larger contributions of electrostatic interaction to the overall stability of complexes. When extrapolated to 1M salt solution, the apparent binding constant is greatly reduced to values comparable to those found for complexes of mononucliotides with amino acids, esters and dipeptides.

Influence of mono- and poly- nucleotides on the yield and rate of peptide synthesis mediated by water soluble (1-ethyl-3(3-dimethyl amino propyl)-carbodiimide)(EADC)

The condensation reaction was carried out with 30mM EDC and 10mm dipeptide in 5mM phosphate buffer at 35°C Fig 3 - shows the time course of condensation of gly-gly and gly-L-tyr at different concentration of ploy C. At 3 hours, a sudden increase in the rate of condensation was observed and at 5 hours the observed decrease is due to the formation of higher oligopeptides. The initial velocity as a function of added poly C is given in the table-2. The initial rate of tetra peptide formation (i.e. coupling constant), is higher for gly-tyr. But the relative change in the

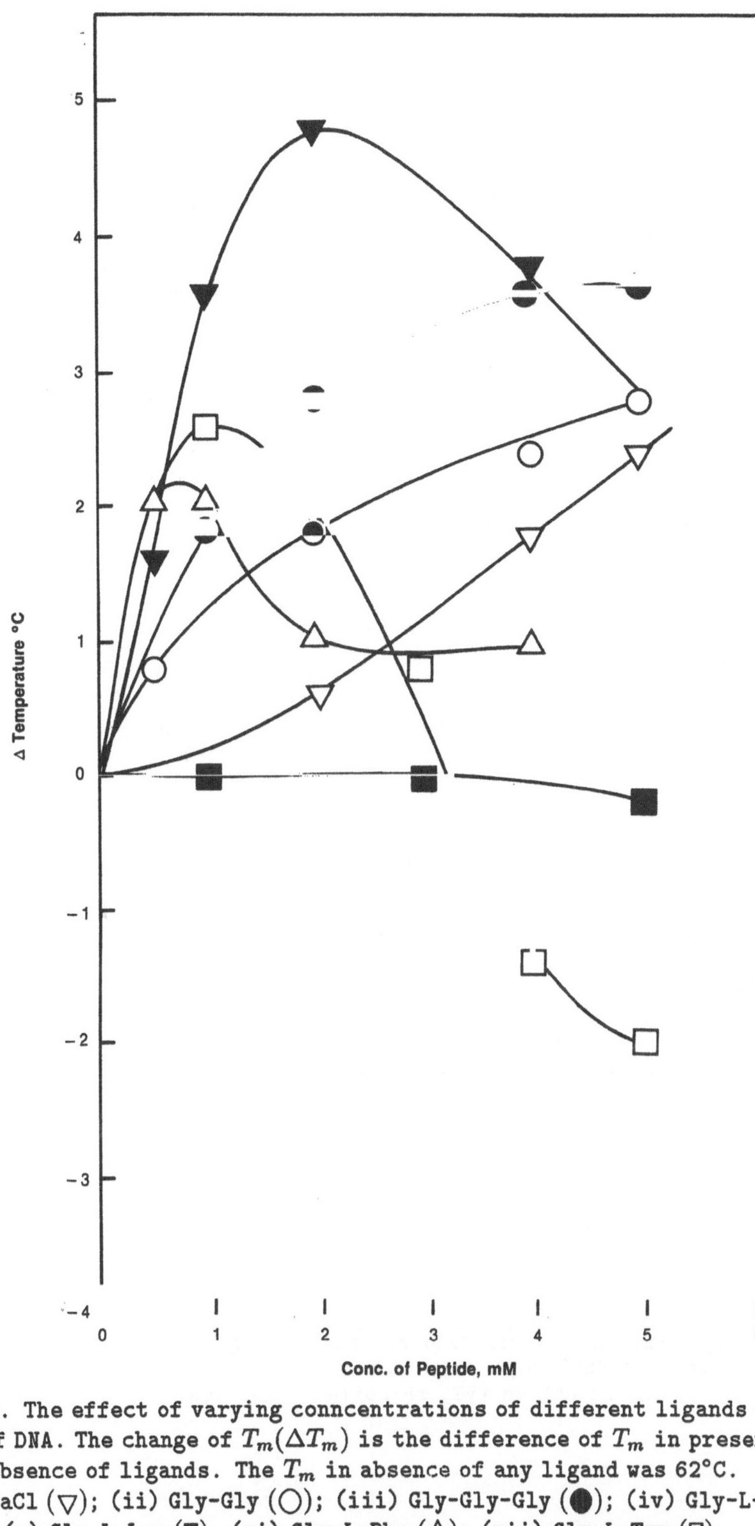

Fig 2. The effect of varying conncentrations of different ligands on the T_m of DNA. The change of $T_m(\Delta T_m)$ is the difference of T_m in presence and absence of ligands. The T_m in absence of any ligand was 62°C. (i) NaCl (▽); (ii) Gly-Gly (○); (iii) Gly-Gly-Gly (●); (iv) Gly-L-Lys (▼); (v) Gly-L-Leu (■); (vi) Gly-L-Phe (△); (vii) Gly-L-Tyr (□).

Table-1. Association constant for the binding of amino acid derivatives to mono-, di- and poly- nucleotides.

Amino acid/Dipeptide	Nucleic Acids					
	AMP	GpC	Poly A	Poly C	PolyA-PolyU	DNA
<u>Meltryl ester of aa[1]</u>						
L-Trp	8.3	-	-	-	-	-
D-Trp	6.5	-	-	-	-	-
L-Met	3.1	-	-	-	-	-
L-Ser	1.2	-	-	-	-	-
L-Phe	5.1	-	11.7	-	-	-
L-Val	2.5	-	-	-	-	-
Gly	1.7	-	-	-	-	-
<u>amides of aa[2]</u>						
Phe					480	
Tyr					750	
Typ					1300	
<u>didpeptide[3]</u>						
gly-gly	-	-	-	2354	-	323(212)
gly-L tyr	-	-	-	-	-	1833(1833)
gly-L tyr	-	276	-	970	-	724(1200)
gly-L phe	-	420	-	-	-	-
gly-D-phe	-	45	-	-	-	-
<u>tripeptide[4]</u>						
$(lys)_3$	-	-	$4.1x10^{6}$ -	- -	-	-
lysDtrp-lys[a]		-	- -	- -	-	$6.2x10^{4}$[a]
lysLtrp-lys[a]	-	-	$1.9x10^{4}$	$0.5x10^{4}$ -	$4.2x10^{4}$	$7.7x10^{4}$[a]
lysphe-lys	-	-	$(2.4x10^{3})$ -	- -	-	-
lys-gly-lys	-	-	$(1.8x10^{3})$ -	- -	-	-

[1]) ref(18) [2]) ref (23) [3]) ref (25) and this study; [4]) ref (10)
[a]) Biochemistry (1985) 24, 4333; numbers in parenthesis in colulmn III refer to $(A)_6$ and in column (VI) to coiled calfthymus.

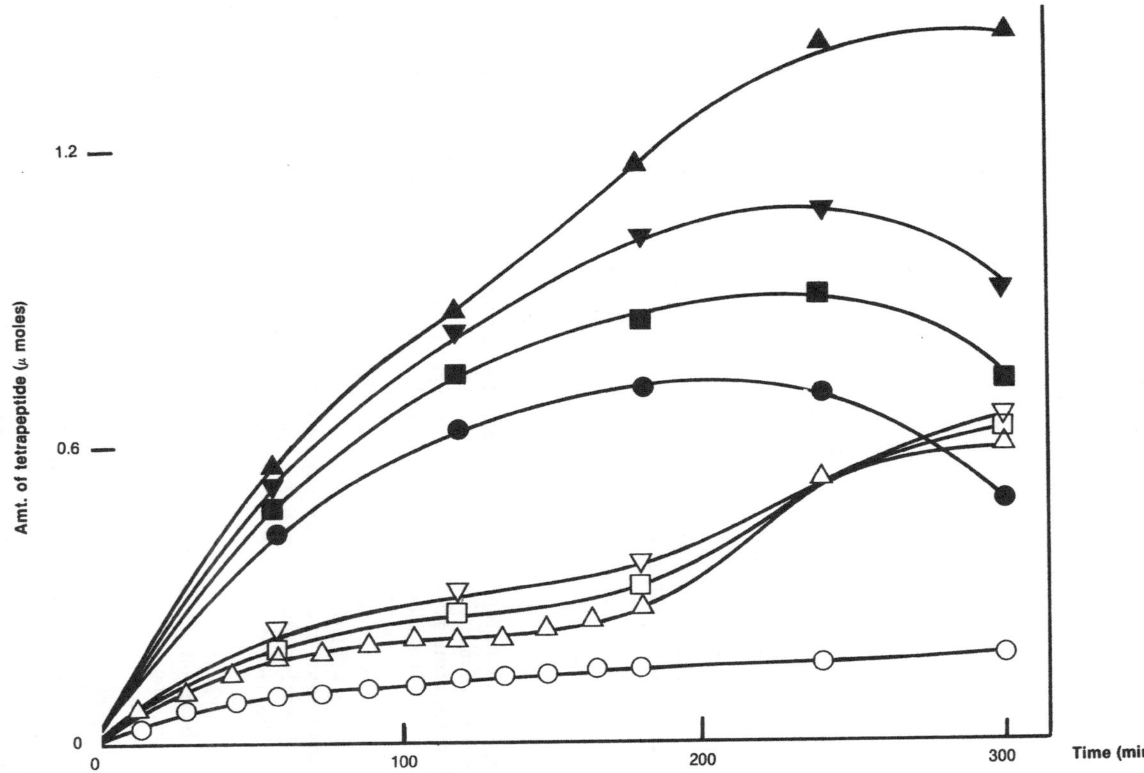

Fig 3. The kinetics of the formation of tetrapeptides during the EDAC mediated condensation of dipeptides in presence of poly(rC) at varying ratios of nucleotide residue:peptide (r). The condensation of Gly-Gly: (i) r = 0.0 (○); (ii) r = 0.33 (△); (iii) r = 0.44 (□); (iv) r = 0.66 (▽). The condensation of Gly-L-Tyr: (i) r = 0.0 (●); (ii) r = 0.33 (■); (iii) r = 0.44 (▼); (iv) r = 0.66 (▲).

Table-2. The Change of Initial Velocity of EDAC mediated Condensation of Dipeptides (calculated from $\Delta c/\Delta t$; t=5 min) in presence of Varying concentration of Poly r(C)

[The initial concentration of all dipeptides was 10mM. The condensation was carried out at 37*C in 5 mM phosphate buffer (pH 7.0)]

Dipeptide	addenda	Ratio(r) nucleotide/ dipeptide	$v^i \times 10^4$ mM sec^{-1} $\Delta c/\Delta t$ t=5 min
Gly-Gly	-	0	0.83
		0.33	1.39
		0.44	1.94
		0.66	2.27
	Gly-Gly-Gly-Gly(10mM)	0	1.40*
Gly-L-Tyr	-	0.33	3.80*
		0	4.72
		0.33	6.22
		0.44	6.66
		0.66	7.00
	Gly-Gly-Gly-Gly(10mM)	0	7.70*
		0.33	18.8 *

*Determined at 1 hr time point to eliminate the complexity arising due to the interference of the tetrapeptide added.

initial rate is found to be same upon addition of exogenous tetraglycine suggesting thereby that the observed increase in the initial rate of tetra-peptide formation is due to the catalysis by tetra-peptide formed in situ. Mononucleotides like the CMP enhanced yield of tetra peptide(25). Fig 4 shows the relative enhancement of tetra-peptide yield in the presence of various templates and fig 5 the correlation between enhancement and the affinity.

It is known that the rate of condensation reaction in water can be affected by influencing any of the following steps: i) the formation of O-acylurea ii) the rearrangement of O-acylurea to N-acyl urea, iii) the aminolysis of O-acylurea by incoming amino acids and peptides iv) hydrolysis of O-acylurea. Due to such complexity the yield are expected to be low as observed with polynucleotide having higher affinity for dipeptide. Since the concentration of a EDAC is rather high, the steps that are affected by polynucleotide are likely to be either formation of O-acylurea or the aminolysis or both. To resolve this problem further, we have estimated the extent of racemisation upon separation of D and L-oligopeptide using matrices to which nucleic acids are coupled(28). The separation of D and L-oligodipeptide are shown in fig 6. The extent of racemisation of gly-L-phe and gly-L-tyr are given in the table-3.

It is seen that the extent of racemisation during condensation of gly-L-phe is about 41% which lies in the range of those obtained by NMR method for condensation of N-Benzoyl-L-phe. with N- Benzoyl-L-lysine(29). The efficiency of different polynucleotides inhibiting racemisation denoted as 'β' is found to exhibit same order of interaction specificity as reflected in the enhancement, thus the extent of racemisation shows a linear relationship with relative enhancement of yield irrespective of the nature of polymeric template(fig. 7). All these taken together suggest that a template specific polymerization takes place. At a given ratio of

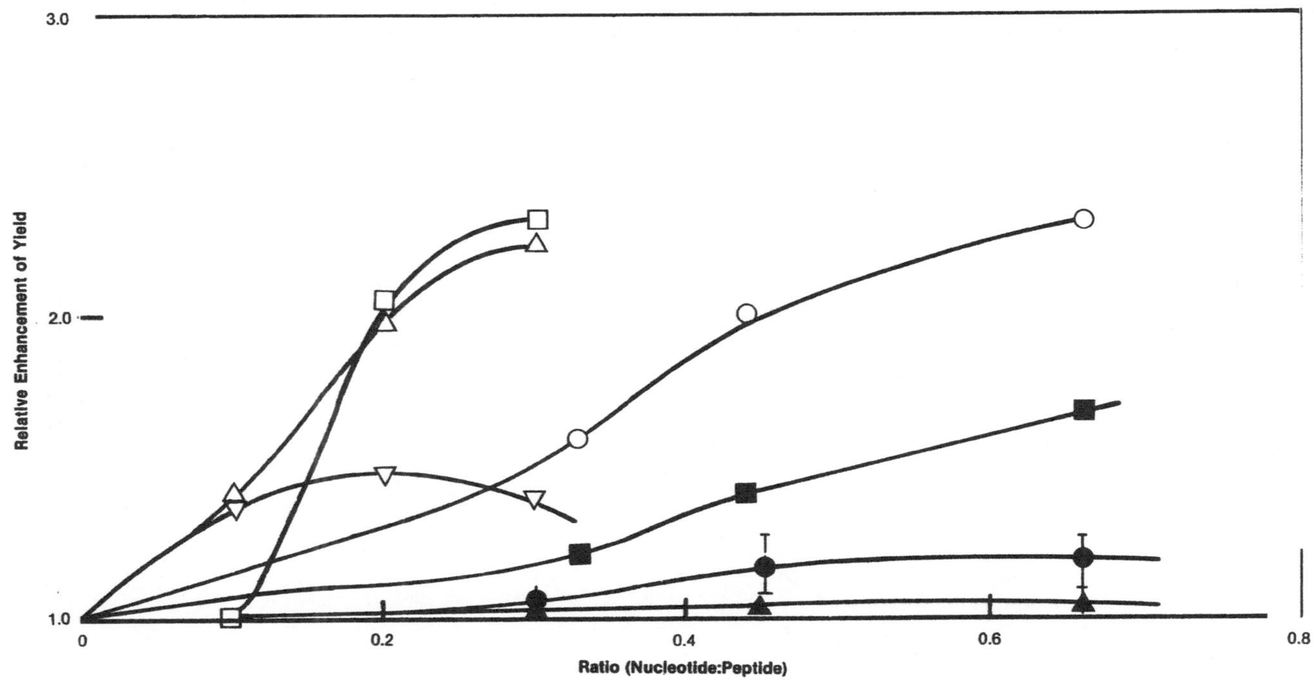

Fig. 4. The relative enhancement of the yield of tetrapeptide due to the EDAC mediated condensation of different dipeptides in presence of different polyelectrolytes at varying ratios of nucleotide residues:peptide. (i) Gly-Gly:polyphosphate (●); (ii) Gly-L-Tyr:polyphosphate (▲); (iii) Gly-Gly:poly(rC) (○); (iv) Gly-L-Tyr:poly(rC) (■); (v) Gly-L-Phe:poly(rA) (△); (vi) Gly-D-Phe:poly(rA) (▽); (vii) Gly-L-Tyr:native DNA (□).

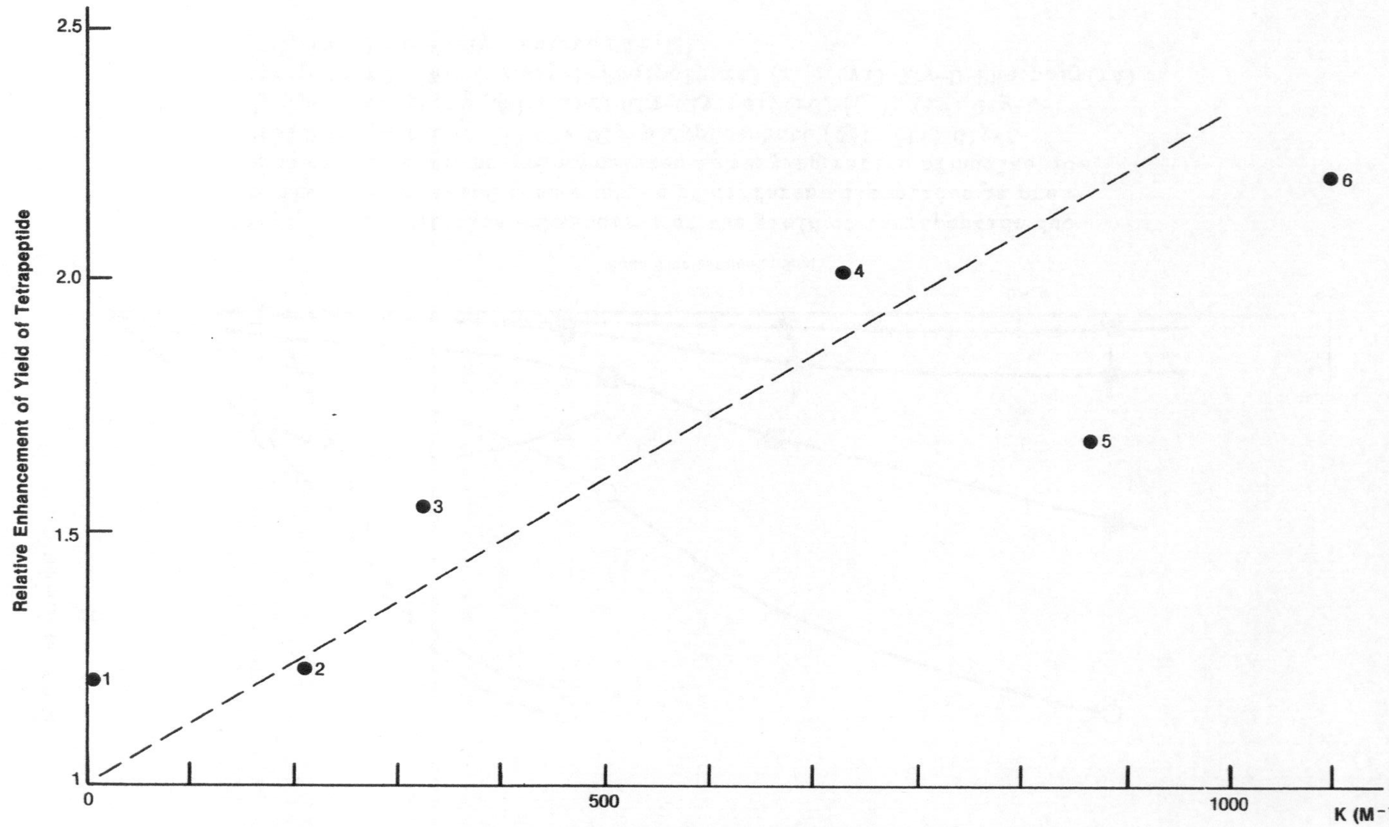

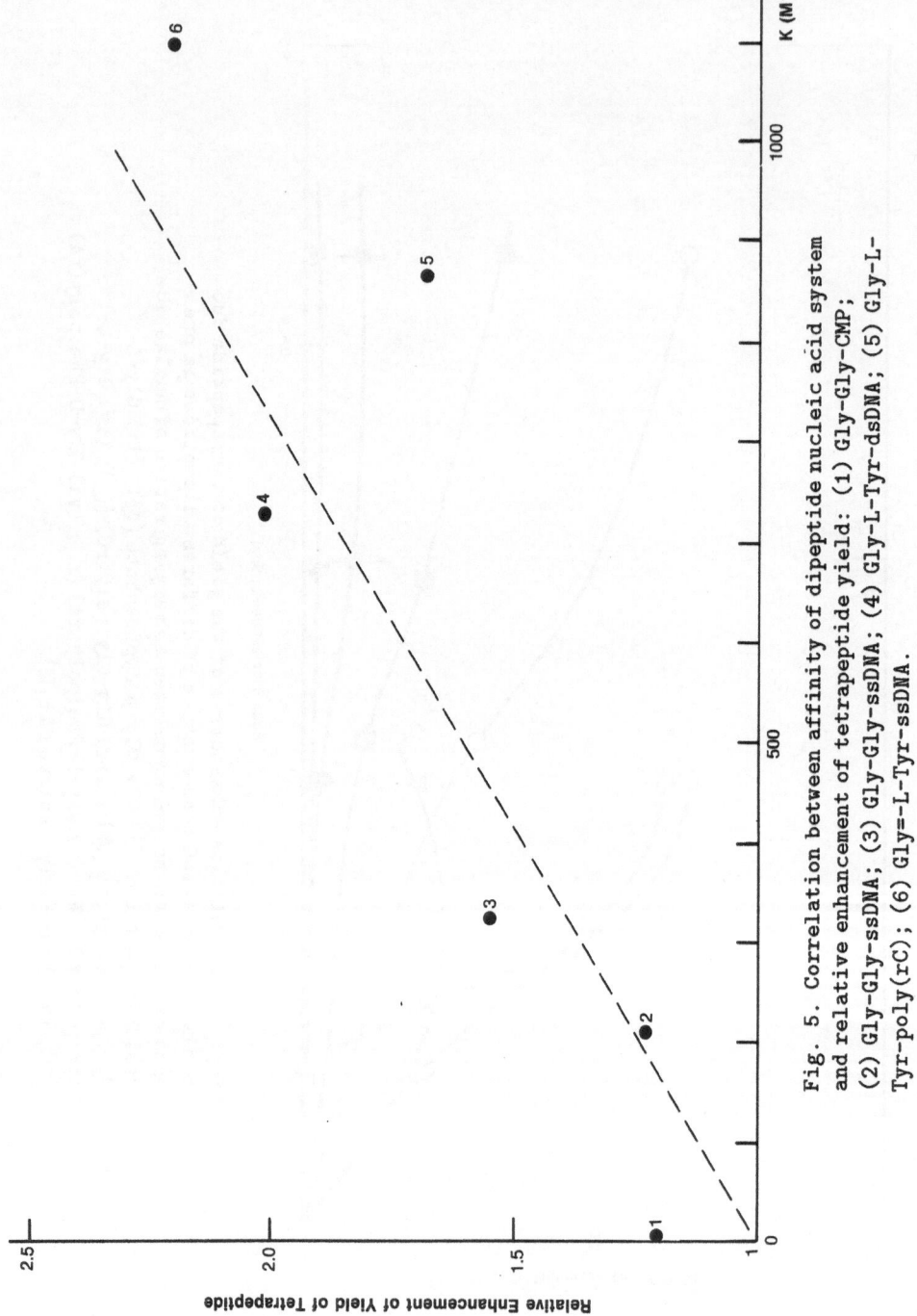

Fig. 5. Correlation between affinity of dipeptide nucleic acid system and relative enhancement of tetrapeptide yield: (1) Gly-Gly-CMP; (2) Gly-Gly-ssDNA; (3) Gly-Gly-ssDNA; (4) Gly-L-Tyr-dsDNA; (5) Gly-L-Tyr-poly(rC); (6) Gly=-L-Tyr-ssDNA.

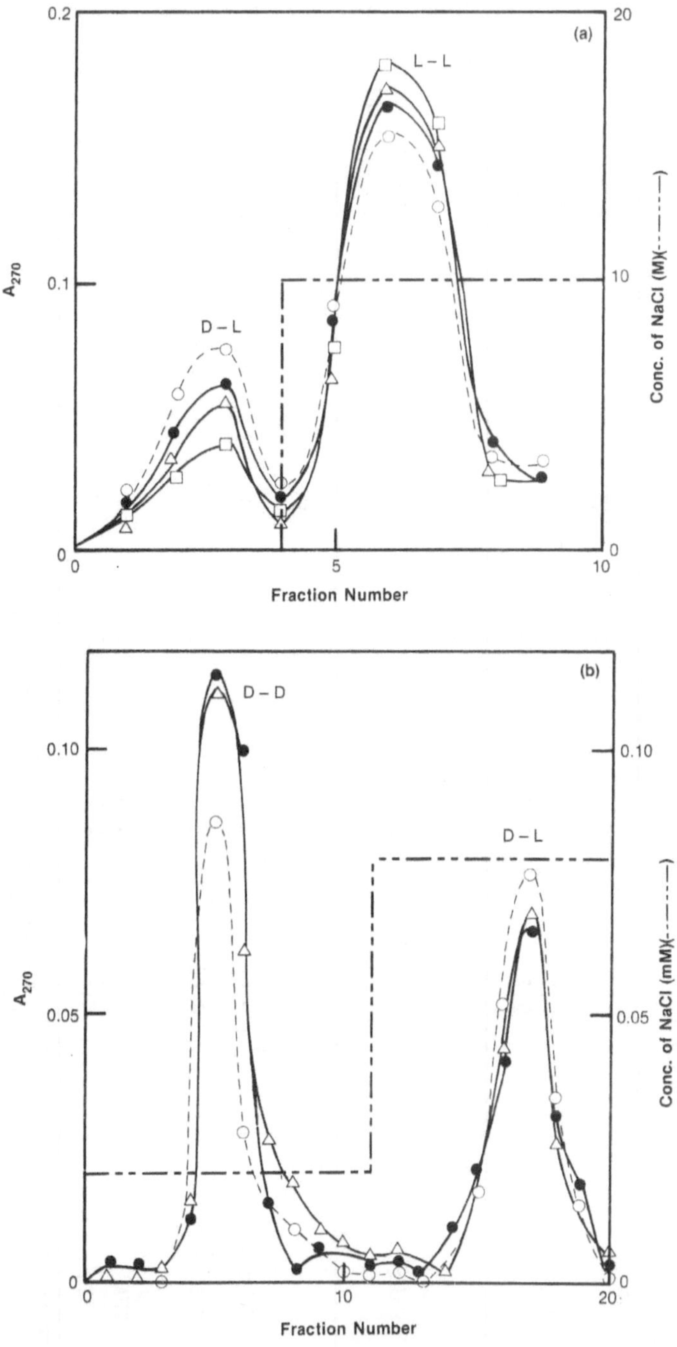

Fig. 6. Elution profiles of the tetrapeptide products obtained from the condensation of Gly-L-Tyr from DNA-sepharose (a); and Gly-D-Phe from (AMP + CMP)-sepharose affinity matrices (b). The peptide condensation reaction was carried out in presence of poly(rA) varying ratios of nucleotide residue:peptide (r). (i) r = 0.0 (○); (ii) r = 0.1 (●); (iii) r = 0.2 (△); (iv) r = 0.3 (□). The A_{270} of the sample loaded = 0.5.

Table 3. Racemization during dipeptide ⟶ tetrapeptide conversion in presence and absense of different polyelectrolytes at varying concentrations.

Peptide	Poly-electrolyte	Ratio of polymer with peptide (r)	Relative enhancement of yield (R)	Percent D-peptide present	Percent L-peptide present	D/L Expt.	D/L Theo β =1)	β*
Gly-D-Phe	-	0	1.0	51	49	1.0	-	
	Poly(rA)	0.1	1.4	56	44	1.27	1.86	
		0.2	1.5	58	42	1.38	2.06	0.72
Gly-L-Phe	-	0	1.0	41	59	0.69	-	
	Poly(rA)	0.2	2.0	27	73	0.37	0.26	
		0.3	2.4	23	77	0.30	0.21	0.87
Gly-L-Try	-	0	1.0	42	58	0.72	-	
	Polyphosphate	0.66	1.1	41	59	0.69	0.62	0.68
	Poly(rC)	0.66	1.4	39	61	0.64	0.43	0.68
	DNA	0.20	2.0	32	68	0.47	0.27	
		0.30	2.6	27	73	0.37	0.19	0.78

*Values obtained from the equation: $\frac{[\text{Unracemized}]}{[\text{Racemized}]} = \frac{(R-1)\ \beta + \alpha}{(R-1)\ (1-\beta) + (1-\alpha)}$

The values of the ratio of [unracemized]/[racemized], R and α are obtained from the data presented above.

Where R = Relative enhancement of yield

α = Fraction unracemized in absence of template

β = Fraction unracemized on the template

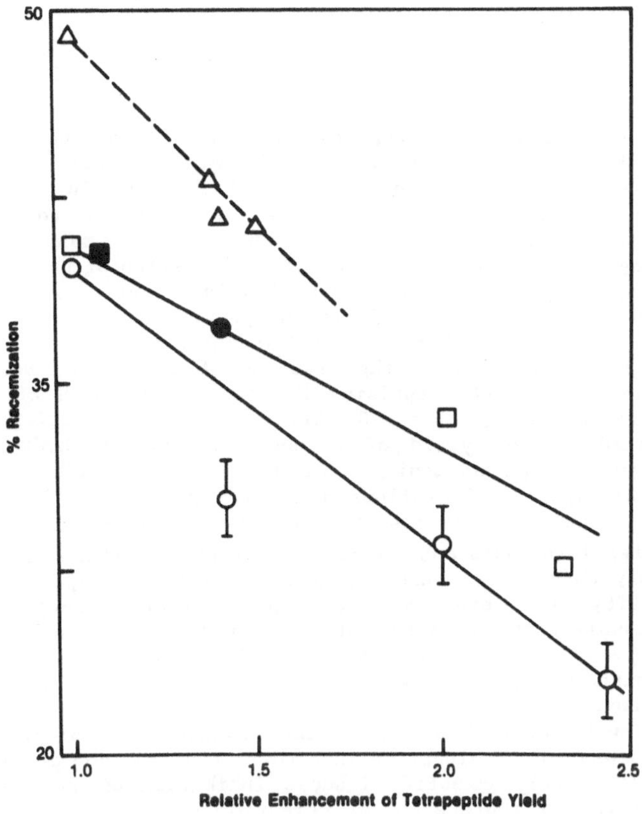

Fig. 7. The correlation of the relative enhancement of tetrapeptide products with the degree of the racemization during the condensation of dipeptides in presence of different polyelectrolytes. (i) Gly-D-Phe:poly(rA) (△); (ii) Gly-L-Tyr:DNA (□); (iii) Gly-L-Tyr:polyphosphate (■); (iv) Gly-L-Tyr:poly(rC) (●); (v) Gly-L-Phe:poly(rA) (○).

nucleotide to peptide the relative enhancement of condensation of different dipeptide is most pronounced in gly-gly: PolyC and gly-L-phe:PolyA system, whereas PolyU has no effect on the condensation of gly-L-phe. This observation of ours demonstrate for the first time that anticodon amino acid and rather than codon amino acid relationship is more pronounced and manifested in template directed nonenzymatic synthesis of peptide. The difference in the intrinsic rate of formation of a peptide bond for different dipeptides suggests that the side chain has a marked effect on the formation of O-acylurea. Due to this intrinsic difference in rates, the specific enhancement of the condensation of any of the dipeptide from equal molar mixture of two different peptides by any polynucleotide could not be monitored.

Discussion

Similar studies have been reported earlier, but no attempt has been made to correlate the enhancement of yield with affinity parameters. Calvin and his workers observed a threefold increase in carbodiimide mediated dimerisation of lysine in the presence of Poly A with nucleotide/peptide ratio of 1:1(30). In contrast, Steinmann & Cole did not observe any appreciable enhancement of dimerisation of phenyl alanine (10mM) in the presence of Poly U with nucleotide: peptide ratio of 0.68 to 2.4 presumably due to lower affinity of phenylalanine to Poly-U(31.) More recently White and Erickson reported an enhancement of the yield of obigomer of glycine in only in the presence of PolyrG and PolyrC(32). This result is not at all surprising in view of the high affinity of glycine oligomer for polyG and polyC. The condensation reaction reported here is limited by low yield of product due to the hydrolysis of condensing agent in aqueous medium. In addition, carbodiimide mediated condensation of amino acids follows a complex path way. Nevertheless these data, together with results presented here would suggest that some nucleic acid template would enhance the synthesis of certain oligopeptide more specificly than other; the extent of enhancement depends not only on the specifity of interaction but also on other factors, (i) rate constant of coupling which in turn depends on the nature of amino side chain and (ii) extent of product catalysis, To examine that enhancement is due to template directed polymerization, the extent of racemisation has also been measured.

It is seen from fig. 7 that the relative enhancement of yield of non enzymatic condensation of dipeptide into tetra peptide runs parallel with the efficiency of polynucleotide induced inhibition of racemisation. This would result in a preferential selection of L-peptide (ie chiral amplification) in view of higher interaction relative to that of any other combination. The specificity is more for amino-acid anticodon pair is also supported by affinity data of amino acid nucleotide.

The recognition of t-RNA by its cognate synthetase is a complex process(33). Both acceptor arm and anticodon loop play a crucial role. Except for ala- and tyr-t-RNA all other transfer RNAs are recognized by their anticodon region. The catalytic efficiency of a cognate enzyme is greatly reduced when anticodon loop was modified(34). Theoretical studies indicate possible association of amino acids with anticodon loop(35). Similarly, chemical modification studies suggest the importance of anticodon region for recognition of the cognate L-amino acid(36). Thus it is tempting to suggest that the stereoselective recognition between amino acid/dipeptide and anticodon nucleotide is the first step in the nucleation of coded translation machinery. At least this is true for homocodonic amino acids like glycine and phenylalanine. It needs to be emphasized that the sequence specific polymersation is unlikely to be all or none process viz the enhancement of gly-tyr condensation product was observed in the presence of PolyC and Calfthymus DNA. Therefore prior information about the relative affinity of other heterocondonic amino acids as well as the rate constant of coupling are required to design experiment relevant to the proposed pathway of origin of coded translation machinery. This is not an impossible task with the advent of biotechnology but certainly not simple and easy. For such a primitive complex to acquire ability to prolificate with high fidelity it should possess rudimentary catalytic efficiency to synthesize any one of the components as suggested by Eigen(37). It is well established that oligopeptide has differential affinity for single and double stranded nucleic acid as we found for dipeptides. The salt sensitive differential affinity could lead to the alteration of local secondary structure of nucleic acid template, and thereby to more complex pattern of enhanced peptide synthesis. Thus it is likely to observe a complex pattern of oligopeptide synthesis if one uses present day t-RNA or single stranded t-RNA like molecules for reasons of belief that t-RNA came first and involved in the establishment of genetic machinery in primitive milieu(38). In short, kinetics of nonenzymatic template directed polymersation of nucleic acids as well as those of amino acid and peptides particularly in the presence of individual t-RNA as reported here needs much more careful study in order to assess whether the catalytic efficiency of template directed synthesized oligopeptide could aid further the evolution of the present day genetic code.

Acknowledgements Financial support from the council of Scientific and Industrial Research (C.S.I.R), India is acknowledged.

References

1. Von Hippel, P(1984) Ann Rev. Biochem 53, 389.
2. Perutska, J. et. al (1988) Proc. Natl Acad. Sci. 85, 6252.
3. Joyce J. and Orgel L. (1988) J. Mol. Biol 202, 607.
4. Joyce J.F. et al (1987) Proc. Natl. Acad. Sci. 84, 4398.
5. Laccy, J. C., Jr. (1988) Proc. Natl. Acad. Sci. 85, 4996.
6. Orgel, L.E. (1986) J. Theo Biol. 123, 127.
7. Dounce, A. L. (1981) J. Theo, Biol 90, 63.
8. Joyce, G. F. (1989) Nature, 338, 217.
9. Lacey, J.C. Jr. and Mullins D.W. Jr. (1983) Origins of life, 13, 3.
10. Helene, C. and Maurizot, J (1981) C.R. C. Critical Review in Biochemsitry, 10,213.
11. Takada, Y. et al (1989) Proc. Natl. Acad. Sci. U.S.A. 86, 439.
12. Helene, C. and Lancelot, G. (1983) Prog. Biophys and Mol. Biol. 39, 1.
13. Ballard, D.G.H. and Bamford, C.H. (1956) Proc, Roy Soc.(London) 236A. 384.
14. Bayer, E. et al (1974) J. Am. Chem. Soc. 96, 7333.
15. Chella, G. and Tan, Y.Y. (1981) Pure and Appl. Chem 53, 627.
16. Steinmann, G. And Cole, M.N. (1968) Fed. Proc. 27, 765.
17. Thomas, P.D. and Podder, S.K. (1978) Febs Letter 96, 90.
18. Reuben, J. and Polk, F. (1980) J. Mol evolution, 15, 103.
19. Khaled, M.A. et al. (1982) Biochem, Biophy Res. Comp. 106, 1426
20. Das Gupta D. and Podder, S.K. (1979) Ind. J. Biochem. Biophy. 16, 316.
21. S.K. Podder and Dasgupta, D. (1980) Ind. J. Biochem Biophy, 17, 417.
22. Das Gupta, D. (1979) Ph.D thesis, Indian Institute of Science, Bangalore, India.
23. Poerschke, D. and Jung, M. (1982) Nuc. Acid Res. 10, 6163.
24. McGhee, I.D. (1976) Biopolymers 15, 1345.

25. Podder, S.K. and Basu, H.S. (1984) Origins of life 14, 477.
26. Gross, E. and Meinhoffer, J. (1979) in the Peptides (Gross E and Meinhoffer, J. ed, Acad. Press) 1, 44.
27. Hoare, D.G. and Koshland, Jr. D.E. (1967), J. Biol. Chem. 242, 2447
28. Basu, H.S. and Podder, S.K. (1982) Ind. J. Biochem & Biophys. 19, 305.
29. Benoiten, N.L. etc. (1980) Int. J. Pep. Prot. Res. 15, 475.
30. Bjornson et al (1974) in the origin of life and evolutionary Biochemistry (Dose K. Fox, S.W. et al) Plenum Press, N.Y.) 21.
31. Steinmann G. and Cole, M.N. (1968) Proc. Natl. Acad. Sci. (U.S.) 58, 735.
32. White, D. H. and Erickson, J.C. (1981) J. Mol. Evolution 17, 19.
33. Schimmel,.P. (1989) Biochemistry, 26, 2747
34. Kisseley, L.E. (1985) Prog. in Nucleic Acid Research and Molecular Biol. 32, 237.
35. Shimizu, M. (1982) J. Mol evolution 18, 297.
36. R. Balasubramanian (1982) TIBS. 7,9.
37. Eigen, M. (1971) Naturwiss 58, 465.
38. Eigen, M. et al (1989) Science 244, 673.

ASYMMETRY AND THE ORIGIN OF LIFE

C. Ponnamperuma[1], Y. Honda[1] and R. Navarro-González[2]

1 Laboratory of Chemical Evolution
Department of Chemistry and Biochemistry
University of Maryland
College Park, Maryland 20742, U.S.A.

2 Instituto de Ciencias Nucleares
U.N.A.M., Circuito Exterior, C.U.
México D.F. 04510, Mexico

Abstract. The origin of life and of asymmetry are integrally related. Several studies have been carried out to obtain a better insight into both problems. This paper summarizes the different accomplishments obtain from our laboratory in the last two decades.

Keywords: Origin of life, Asymmetry, Enantiomer, Meteorite, Prebiotic synthesis, Clay, Amino acid, Model peptide, Nucleotide, Genetic code.

1. Introduction

Modern scenarios of the origin of life on Earth are based on ideas generated independently by Oparin (1924) and Haldane (1929). According to these authors the origin of life was preceded by abiotic syntheses and accumulation of organic compounds of increasing complexity on the primitive Earth. These compounds were essential for the origin and evolution of phase-separated systems that eventually led to the emergence of the first populations of living cells (Oparin, 1924, 1938, 1957, 1972; and Haldane, 1929).

It is very clear that the origin of life itself is integrally related to the origin of optical activity (Ponnamperuma, 1974). However, it is not yet known if optical activity originated during the synthesis of biomolecules, or during the evolution of phase-separated systems, or during the early biological evolution.

Several methods have been applied to understand the origins of life and of optical activity. In our laboratory we have studied these problems through the analytical and synthetic approaches. In the analytical approach, we have been able to go back in time and look at the records of ancient rocks and/or sediments from the Earth and from extraterrestrial bodies such as the Moon, Mars, and

Symmetries in Science IV
Edited by B. Gruber and J. H. Yopp
Plenum Press, New York, 1990

meteorites. Our synthetic approach has counted of the studies related to: i) the synthesis of small molecules by various forms of energy under plausible primitive Earth conditions; ii) the interaction among small molecules with inorganic matrices; and iii) the association between amino acids and nucleotides.

In this paper we will describe our experimental results and attempt to relate them to other studies which will help us to understand the origins of life and of asymmetry.

2. The search for organic compounds in the solar system

Of paramount importance in the understanding of the processes that led to the beginning of life is the information that can be gathered from the study of organic matter in the solar system. In this regard we have searched for evidence of chemical evolutionary processes in the Moon, Mars, and meteorites.

Analysis of lunar samples (Chang et al., 1971; Chang and Kvenvolden, 1972) and martian rigolith (Biemann et al., 1976 and 1977; Horowitz et al., 1976) showed that organic molecules were present in very low abundance. No attempt was made to examine the chirality of the molecules present.

Meteorites are sources of extraterrestrial material amenable to laboratory study on Earth. In 1969 the Murchison meteorite which landed in Australia was analyzed within days. The first analysis established unequivocally that six amino acids commonly found in proteins, and eight which do not occur in natural proteins, were present in the meteorite (Kvenvolden et al., 1971). The amino acids, alanine, aspartic acid, glutamic acid, proline, and valine, were present in approximately equal amounts of the D and L-isomeres (Figure 1.a). Further analysis clearly established this for a large number of amino acids. Similar results have now been obtained with other meteorites such as the Murray (Figure 1.b, Lawless et al., 1971), the Mighei, the Alan Hills, and the Yamato (Ponnamperuma, 1972). Figure 1

The presence of non-biological amino acids and almost equal amounts of the D and L-enantiomers minimizes the possibility of terrestrial contamination, and suggests a possible extraterrestrial origin. The question immediately arises as to their origin. Two possibilities present themselves: i) these amino acids were existing at some period of time in the meteorite in one stereo-isomeric form and were then racemized in the course of time; or ii) the two forms were always present in nearly equal amounts (Kvenvolden et al., 1971). The possibility that an extraterrestrial life was responsible for D or L-amino acids initially cannot be rejected entirely. On the other hand, the production of either D or L-isomer by prebiotic processes and subsequent racemization may be considered unlikely, since no well-defined evidence is yet available for abiotic processes that would produce one form rather than the other. Therefore it is more likely that amino acids in meteorites were produced in racemic mixtures by abiotic processes.

3. Prebiotic synthesis

We have synthesized a number of prebiotically important molecules such as hydrocarbons, aldehydes (Honda, et al., 1989), carboxylic

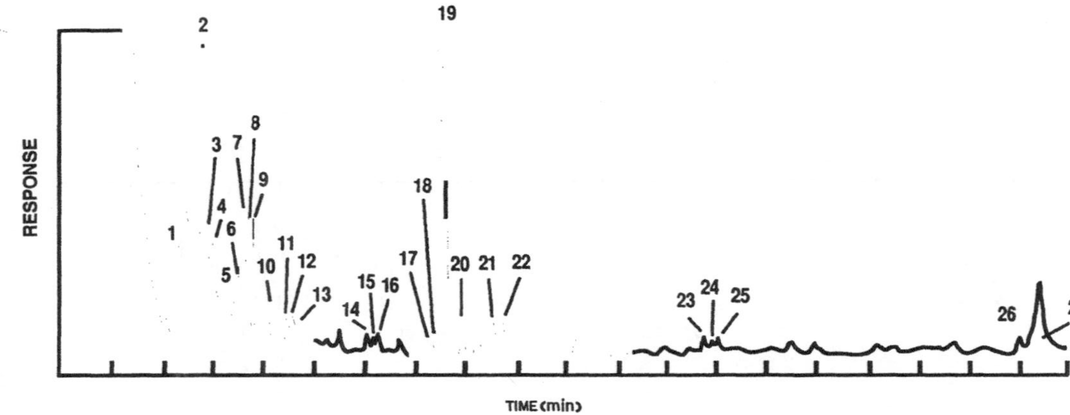

Fig. 1a (above) and Fig. 1b (next page). Gas chromatograms of N-trifluoroacetyl-D-2-butyl esters of amino acids in acid-hydrolyzed aqueous extracts of Murchison (a) and Murray (b) meteorites (Ponnamperuma, 1972). Identification for the samples is as follows: (1) (DL?)-isovaline; (2) α-amino-isobutyric acid; (3) D-valine; (4) L-valine; (5) (DL?)-N-methylalanine (?); (6) D-amino-n-butyric acid; (7) D-α-alanine; (8) L-α-amino-n-butyric acid; (9) L-α-alanine; (10) N-methylglycine; (11) N-ethylglycine (?); (12) D-norvaline (?); (13) L-norvaline (?); (14) D-β-aminoisobutyric acid (?); (15) L-β-aminoisobutyric acid (?); (16) (DL?)-β-amino-n-butyric acid; (17) D-pipecolic acid; (18) L-pipecolic acid; (19) glycine; (20) β-alanine; (21) D-proline; (22) L-proline; (23) γ-amino-n-butyric acid; (24) D-aspartic acid; (25) L-aspartic acid; (26) D-glutamic acid; (27) L-glutamic acid; (masked by an unidentified peak).

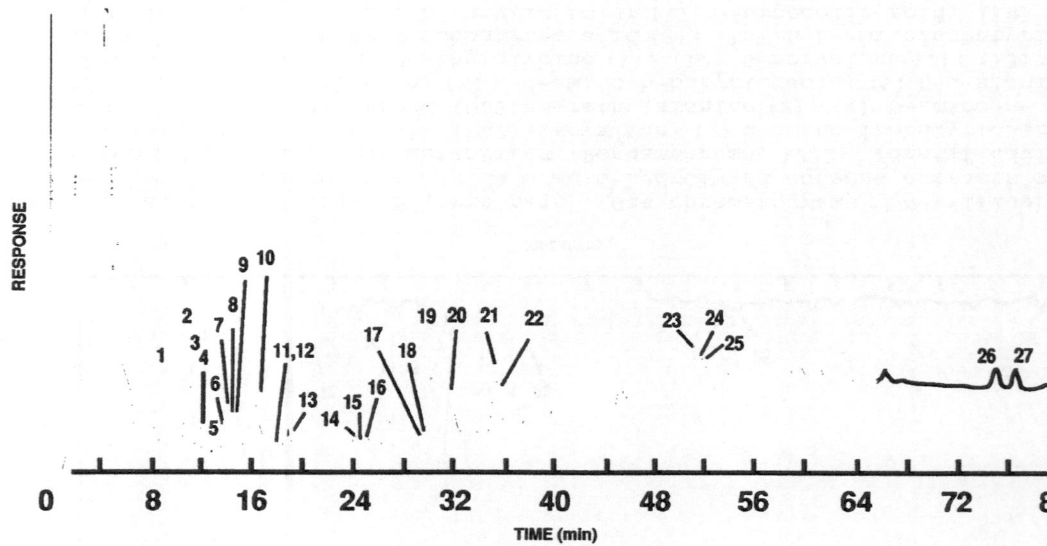

Fig. 1b. See figure caption on preceeding page, Fig. 1a.

acids (Negrón-Mendoza, et al., 1983; Honda, et al., 1989), amino acids (Kobayashi and Ponnamperuma, 1985), and nucleobases (Ponnamperuma et al., 1963a,b; Kobayashi et al., 1986) in experiments that simulate possible environments of the primitive Earth. All possible isomers are usually produced in prebiotic synthesis, e.g., C_5: 2- and 3-methylpentene; C_6: 2,4- and 3,4-dimethylhexane (Ponnamperuma and Woeller, 1964).

Enantiomeric analysis of amino acids produced in electric discharge experiments was reported in the presence or absence of clay (Shimoyama et al, 1978). In this study the relative abundance of enantiomers was determined by their peak areas in the gas chromatograms. The results show that D and L-enantiomers were formed in equal amounts (Figure 2). The ratios varied from 48% to 52% for alanine and α-amino-n-butyric acid. This variation was also obtained for the same sample when D and L-stationary phases were used. Discrepancies of this magnitude are frequently observed for authentic D and L-amino acid samples. Therefore, it is reasonable to conclude that the amino acids are produced in equal amounts. The presence or absence of clay in the systems did not affect in the enantiomeric synthesis of amino acids. Figure 2

Further investigations are needed to extend other chemical systems and energy sources to examine any selective synthesis of L or D-enantiomers. Nonetheless based on our current results, it seems likely that wherever an organic compound with an asymmetric center is formed almost equal amounts of L and D-isomers appear to be present.

4. Selective Adsorption of Enantiomers in Mineral Matrices

One of most interesting possible roles of clays in chemical evolution which was suggested by Bernal in 1951 is the preferential adsorption of certain biological monomers. Many investigators have studied the adsorption of biologically important organic compounds in clays (Ponnamperuma et al, 1982). The possible selective adsorption of optical isomers by clays has also been of interest in the study of the origin of asymmetric molecules. In order to determine whether clays could have played a part in selection process of L amino acids over D forms in prebiotic times, the adsorption of several amino acids in Na-montmorillonite was studied at various pH's, see Table 1 (Friebele et al., 1980).

Gas chromatography analyses showed that the L-enantiomeric form of amino acids was in 0.5-2.0% more adsorbed in the clay than the D-form. However enantiomeric analyses of the same samples by amino acid analyzer yielded results which were not consistent in all cases with gas chromatographic data (Table 1). The results suggest that apparently clays have no significant ability to sort out L and D-amino acids under the experimental conditions. Our findings demonstrated that clays may not have played an important role in the enantiomeric selection of amino acids in the process of chemical evolution.

5. Interaction between amino acids and nucleotides

The origin of nucleic acid-directed protein synthesis is one of the principal problems in the study of the origin of living systems. In contemporary organisms, the genetic information recorded in DNA is first transcribed into mRNA, and the nucleotide sequence of mRNA is then translated into the amino acid sequence of proteins. Each group of three consecutive nucleotides defined as a codon, in the

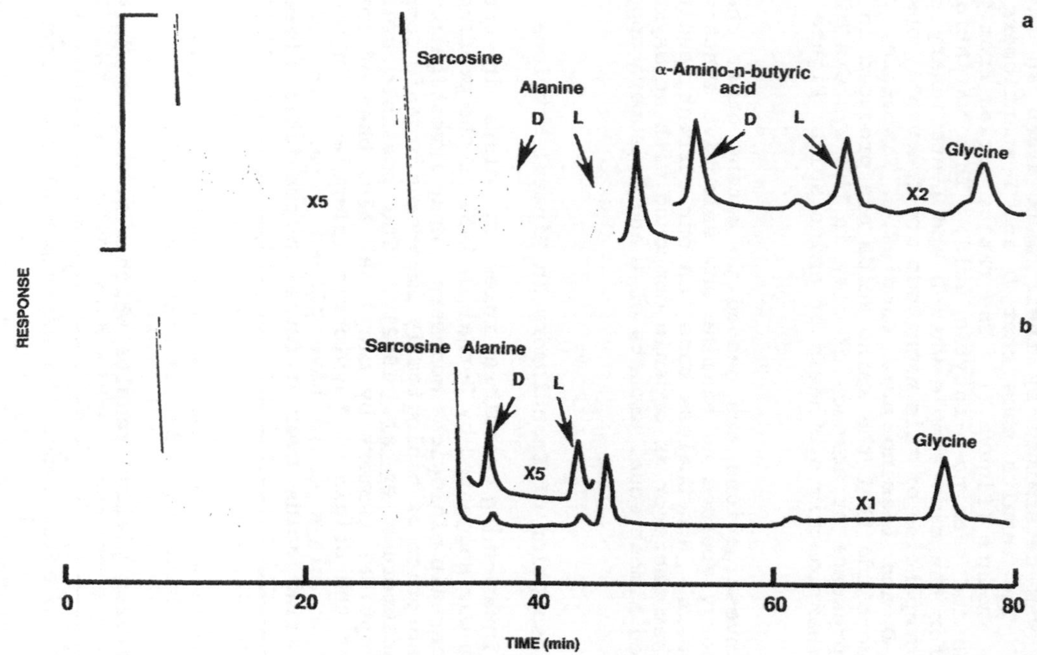

Fig. 2. Gas chromatogram of amino acid enantiomers produced in the electric discharge from a simulated atmosphere-hydrosphere environment in the presence (a) or absence (b) of Na-montmorillonite (Shimoyama, et al., 1978).

mRNA, is read as a signal for the incorporation of a particular amino acid. To achieve this, amino acids are attached to a special class of adaptor molecules, the tRNAs, each of which carry a special trinucleotide sequence known as the anticodon, by synthetase. The question to be answered concerns the physical basis by which nucleic acid components could contain information regarding amino acids, thereby originating the process that led to the evolution of this current elaborated system of coding and translation.

TABLE 1. D:L Ratio of Amino Acids Adsorbed on Na-Montmorillonite as Determined by Gas-Liquid Chromatography[1].

Amino Acid	pH		
	3	7	10
α-Alanine	1.022	0.976	0.953
		1.001*	0.983*
α-Aminobutyric acid	1.005	0.996	0.925
Norvaline	0.986	0.979	0.986
Valine	0.977	0.996	1.015

[1]Friebele, et al., 1981.

*Analysis determined by amino acid analyzer.

Speculations about the origin of the genetic code began even before the code was deciphered (Pauling and Delbruck, 1950; Gamow, 1954; Nirenberg and Mattaei, 1961; Leder and Nirenberg, 1964). At one extreme of these speculations is the direct interaction theory which states that the codon assignments are based on preferential physical and chemical interactions between amino acids and component nucleotides of the genetic code. This idea is amenable to experimental verification as it may be presumed that traces of those early selective interactions are even now discernible, though they may not play any role in the contemporary system.

Since the deciphering of the genetic code, stereochemical models for the origin of the genetic code have fallen primarily into two classes, that is, amino acid-codon and amino acid-anticodon interaction models (Ponnamperuma and Hobish, 1984). Studies on the association of amino acids and nucleotides have been done in our laboratory to understand the origin of the codon assignments. Measurements of the changes in the chemical shifts by NMR in an amino acid-nucleotide mixture allows quantitative determinations of the binding process.

The association of homocodonic amino acids (L-Phe, D-Phe, L-Lys, L-Pro, and Gly) with nucleotide 5'-monophosphates (AMP, UMP, GMP, and CMP) was analyzed. L-Homocodonic amino acids have higher association constants with anticodonic nucleotides than with codonic nucleotides. In addition, anticodonic nucleotides have low association constants with D-homocodonic amino acids than with L-homocodonic amino acids (Senaratne, 1986). These results provide a possible physical and chemical bases for the origin of the genetic code. Further work is needed to look at the interactions between L-nucleotides and D-amino acids to learn more of the roll of asymmetry in the origin of genetic code of living organisms.

6. Synthesis and properties of asymmetric biomolecules

We have been interested in examining the effect of enantiomeric inversion of amino acid residues on protein structure, and have formed a synthetic fragment complex whose properties provide variable insights into the relationships between structure and function of polypeptides of biological interest. In this respect, one of most well-characterized systems is bovine pancreatic ribonuclease S, a stable and functionally active complex consisting of S-peptide (residues 1-20) and S-protein (residues 21-124).

An all-L model S-peptide, H-Ala-Glu-Ala_4-Lys-Phe-Ala-Arg-Ala-His-Met-Ala_2-OH, have been previously synthesized and shown to form a stable complex with S-protein, exhibiting 36% of the specific activity of fully native ribonuclease S against the substrate, cytidine 2',3'-cyclic monophosphate, D-cCMP, (Komoriya and Chaiken, 1982). In order to examine the effect of a defined enantiomeric sequence on protein structure, the all-D model S-peptide was synthesized.

In the presence of S-protein, the all-L S-peptide has hydrolytic activity against the substrate D-cCMP, whereas the all-D peptide was inactive. Addition of the all-D peptide to mixtures of L-peptide and S-protein did not lead to inhibition of enzymatic activity. These results indicate lack of binding of D-peptide to S-protein to produce either an active or inactive species (Corigliano-Murphy et al., 1985). To get a better insight into the catalytic activity of an all-D model ribonuclease S, we would need to synthesize an all-D model S-protein and to assay its activity against L-cCMP as substrate.

7. Conclusions

While considering the problem of the origin of life, it is necessary to give special attention to the origin of asymmetry. Our attempts to understand both problems have been the analytical and synthetic approaches. The analysis of meteorites clearly indicates the presence of racemic mixture of amino acids. While several possibilities may explain the existence of equal amounts of D and L-forms; nevertheless it is more likely that both forms were produced by abiotic processes. This suggestion is supported not only by our experimental evidence on simulated prebiotic syntheses but also by the examination of selective adsorption of enantiomers in mineral matrices. Studies on the interaction between amino acids and nucleotides have provided possible physical and chemical bases for the origin of the genetic code. These results indicate that anticodonic nucleotides have higher association constants with L-homocodonic amino acids than with the D-forms. Further work is however needed to investigate the interactions between L-nucleotides and D-amino acids. We are in addition examining the effect of enantiomeric inversion of amino acid residues on protein structure and function, and have synthesized an all-D model enzyme fragment. To obtain an insight into catalytic activity of an enantiomeric inverted enzyme it is still needed to synthesize the other fragment of the enzyme complex. This would perhaps help us to understand some aspects of the origin of asymmetry.

References

Bernal, J. D.:1951, The Physical Basis of Life., Routledge and Kegan Paul, London, p.34.

Biemann, K., Oró , J., Toulmin, P., E., Orgel, L. E., Nier, A. O.,Anderson, D. M., Simmonds, P. G., Flory, D., Diaz, A. V., Rushneck, D. R., and Biller, J. A.:1976, Search for Organic and Volatile Inorganic Compounds in Two Surface Samples from the Chryse Planitia Region of Mars., Science, 194, p.72.

Biemann, K., Oró, J., Toulmin, P., E., Orgel, L. E., Nier, A. O., Anderson, D. M., Simmonds, P. G., Flory, D., Diaz, A. V., Rushneck, D. R., Biller, J. A. and Lafleur, A. L.:1977, The Search for Organic Substances and Inorganic Volatile Compounds in the Surface of Mars., J. Geophys. Res., 82, p.4641.

Chang, S. K., Kvenvolden, K., Lawless, J., and Ponnamperuma, C.:1971, Carbon, Carbides, and Methane in an Apollo 12 Sample., Science, 171, p.474.

Chang, S. and Kvenvolden, K. A.:1972, Distribution and Significance of Carbon Compounds on the Moon., in Exobiology (Ponnamperuma, C. ed.), North-Holland, Amsterdam, p.400.

Corigliano-Murphy, M. A., Liang, X., Ponnamperuma, C., Dalzoppo, D., Fontana, A., Kanmera, T. and Chaiken, I. M.:1985, Synthesis and Properties of an All-D Model Ribonuclease S-Peptide., Int. J. Peptide. Protein Res., 25, p.225.

Friebele, E., Shimoyama, A., and Ponnamperuma, C.:1980, Adsorption of Protein and Non-Protein Amino Acids on a Clay Mineral: A Possible Role of Selection in Chemical Evolution., J. Mol. Evol., 16, p.269.

Friebele, E., Shimoyama, A., Hare, P. E., and Ponnamperuma, C.:1981, Adsorption of Amino Acid Entantiomers by Na-Montmorillonite., Origins of Life, 11, p.173.

Gamow, G.:1954, Possible Relation between Deoxyribonucleic Acid and Protein Structure, Nature, 173, p.318.

Gehrke, C. W., Zumwalt, R. W., Kuo, K., Ponnamperuma, C., and Shimoyama, A.:1987, Search for Amino Acids in Apollo Returned Lunar Soil., in Amino Acid Analysis by Gas Chromatography (Gehrke,C. W., Kuo, K. C. T., and Zumwalt, R. W. eds.), CRC Press, Boca Raton, Florida, Vol. II, p.151.

Haldane, J. B. S.:1929, The Origin of Life., The Rationalist Annual, Vol. 148, p.3, Republished in The Origin of Life, 1967, Bernal, J. D., Weidenfeld and Nicolson, London, p.242.

Honda, Y., Navarro-González, R. and Ponnamperuma, C.:1989, A Quantitative Assay of Biologically Important Compounds in Simulated Primitive Earth Experiments. Advances in Space Research, 9, p.(6)63.

Horowitz, N. H., Hobby, G. L., and Hubbard, J. S.:1976, The Viking Carbon Assimilation Experiments: Interim Report., Science, 194, p.1321.

Kobayashi, K. and Ponnamperuma, C.: 1985, Trace Elements in Chemical Evolution II: Synthesis of Amino Acids under Simulated Primitive Earth Conditions in the Presence of Trace Elements., Origins of Life, 16, p.57.

Kobayashi, K., Hua, L.-L., Gehrke, C. W., Gerhardt, K. O., and Ponnamperuma, C.:1986, Abiotic Synthesis of Nucleic Acid Bases by Electric Discharge in a Simulated Primitive Atmosphere., Origins of Life, 16, p.299.

Komoriya, A. and Chaiken, I. M.:1982, Sequence Modeling Using Semisynthetic Ribonuclease S., J. Biol. Chem., 257, p.2599.

Kvenvolden, K., Lawless, J., Pering, K., Peterson, E., Flores, J., Ponnamperuma, C., Kaplan, I. R., and Moore, C.:1970, Evidence for Extraterrestrial Amino Acids and Hydrocarbons in the Murchison Meteorite., Nature, 228, p.923.

Kvenvolden, K., Lawless, J. G., and Ponnamperuma, C.:1971, Nonprotein Amino Acids in Murchison Meteorite., Proc. Natl. Acad. Sci. U.S., 68, p.486.

Lawless, J. G., Kvenvolden, K. A., Peterson, E., Ponnamperuma, C., and Moore, C.:1972, Amino Acids Indigenous to the Murray Meteorite, Scince, 173, p.626.

Leder, P. and Nirenberg. M. W.:1964, RNA Codewords and Protein Synthesis, II. Nucleotide Sequence of a Valine RNA Codeword., Proc. Natl. Acad. Sci. U.S., 52, p.420.

Negrón-Mendoza, A., Draganic, Z.D., Navarro-González, R. and Draganic, I.G.: 1983, Aldehydes, Ketones, and Carboxylic Acids Formed Radiolytically in Aqueous Solutions of Cyanides and Simple Nitriles, Rad. Res. 95, p.248.

Nirenberg, M. W. and Mattaei, J. H.:1961, The Dependence of Cell-Free Protein Synthesis in E. Coli upon Naturally Occurring or Synthetic Polyribonucleotides., Proc. Natl. Acad. Sci. U.S., 47, p.1588.

Oparin, A.I.: 1924, Proiskhozhdenie zhizny, Moscow. Izd. Moskovshii Rabochii, Translation by Bernal, J. D., 1967, The Origin of Life, Weidenfeld and Nicolson, London, p.199.

Oparin, A.I.: 1938, The Origin of Life, Dover, New York.

Oparin, A.I.: 1957, The Origin of Life on the Earth, 3rd Ed. Oliver& Boyd, London.

Oparin, A.I.: 1972, The Nature of Origin of Life., in Comparative Planetology (Ponnamperuma, C., ed.), Academic Press, New York, p.1.

Pauling, L. and Delbruck, M.:1940, Nature of the Intermolecular Forces Operative in Biological Processes, Science, 92, p.77.

Ponnamperuma, C., Lemmon, R. M., Mariner, R., and Calvin, M.:1963a, Formation of Adenine by Electron Irradiation of Methane, Ammonia, and Water, Proc. Natl. Acad. Sci. U.S., 49, p.737.

Ponnamperuma, C., Young, R. S., and Munoz, E.:1963b, The Formation of Guanine during the Thermal Polymerization of Amino Acids., Fed. Proc., 22, p.479.

Ponnamperuma, C. and Woeller, F.:1964, Differences in Charactor of C_6 to C_9 Hydrocarbons from Gaseous Methane in Low-Frequency Electric Discharges, Nature, 203, p.272.

Ponnamperuma, C.:1972, Organic Compounds in the Murchison Meteorite., in Ann. N.Y. Acad. Sci.(Johnson, F. M. ed.), 194, p.56.

Ponnamperuma, C.:1974, Optical Activity and Origin of Life., in International Symposium on Generation and Amplification of Asymetry in Chemical Systems (Thiemann, W. ed.), Kernforschungsanlage JÜlich, West Germany, p.131.

Ponnamperuma, C., Shimoyama, A., and Friebele, E.:1982, Clay and the Origin of Life., Origins of Life, 12, p.9.

Ponnamperuma, C., and Hobish, M. K.:1984, The Stereochemical Approach to Studies of the Origin of the Genetic Code, in Molecular Evolution and Protobiology (Matsuno, K., Dose, K., Harada, K., and Rohlfing, L. eds.)., Plenum Publishing Co., New York, p.295.

Senaratne, S. M. D. N.:1986, Direct Interactions between Amino Acids and Nucleotides as a Possible Explanation for the Origin of the Genetic Code., Doctoral disertation, University of Maryland.

Shimoyama, A., Blair, N., and Ponnamperuma, C.:1978, Synthesis of Amino Acids under Primitive Earth Conditions in Presence of Clay., in Origin of Life (Noda, H. ed.), Center for Academic Publications, Tokyo, Japan, p.95.

SOME IDEAS ON EVOLUTION OF SYMMETRY BREAKING OF BIOSPHERE ON EARTH

"Symposium Symmetry IV, Schloß Hofen, near Bregenz, Austria"

by W. Thiemann, Universität Bremen

INTRODUCTION TO THE PROBLEM

Nature offers us spectacular demonstration of symmetry breaking in the architecture of the molecular building blocks of biological organisms. It has been pointed out by numerous authors for over a century that the monomer aminoacids (making up the proteins) as well as the sugars (making up the back bone of DNA and RNA) are almost exclusively of the strictly dissymmetric left-resp. right-handedness which has disturbed the minds of symmetry-obsessed chemists and physicists. This observation violates not only the "esthetics" of science but equally experimental evidence where organic chemists always synthetize in laboratory only the racemic compounds, i.e. a stoichiometric mixture of both the right and the left handed molecular in equal amounts.

From the well-documented laboratory experience it was concluded that such "enantiomer" chemical molecular entities were undistinguishable by the usual scalar properties such as the enthalpies of formation, various bond strengths, optical properties, melting points, etc.; the only way to distinguish the right- from a left-handed molecule is based on vectorial properties such as screw directions in which the individual atoms are hooked up in relation to each other, which again is manifested experimentally in the way linear polarized light is rotated or circular polarized light is absorbed in relation to its counterclockwise polarized light.

But then - why prefers nature only the one sort of building elements out of the set of each two mirror images for making biology going ? - A better question - or problem definition - would be not to ask why but how had nature managed to do so in the process of prebiotic or early biotic evolution on Earth.

Symmetries in Science IV
Edited by B. Gruber and J. H. Yopp
Plenum Press, New York, 1990

The "Why" can be answered rather quickly: Because the very precise functionality and effiency of re-production, mutation, metabolism, diversity of species, etc. would not have been accomplished on the base of a symmetric molecular architecture: Complex conformations like double helices and related structures had been out of scope for symmetric molecules, asymmetry on the molecular level was an absolute necessity for the success of evolution from chaotic chemistry to well ordered and highly complex biological organisms.

So we are left with the question, how did chirality, (= the choice of one enantiomer) in biology arise in evolution. And even if we came close to answering the how, what was the driving force to push the decision in just the one direction (namely the L-aminoacids, the D-sugars) and not the other ?

THE THEORIES

Essentially there are schools of thoughts in the literature which are fundamentally opposed to each other: I would like to call them for the sake of simplicity the <u>Randomists</u> as opposed to the <u>Determinists</u>. Both groups claim to have the better convincing arguments on their side.

<u>A) The Randomists</u>

Randomists argue that - in spite of the apparent disagreement with other laws of nature, which commonly allow for a symmetric architecture of universe - the fact that the basic chemical elemts are strongly symmetry-violated, can be understood rather simply as an "intelligent" trick to reach a higher complexity, larger negentropy manifested in the ingenious blue prints of biological organism, both of the unicellular and multicellular device ! And -furtheron argued- the choice for a left-handed amino acid world (and consequently a right-handed ribose world) was a purely random accident, no higher master plan was needed to determine this decision met at an early stage of evolution.
A plausible model for random accident event could look like the following:

Similar to laboratory experience let us assume a small pond filled with an aqueous solution of supersaturated free aminoacids of the racemic type, i.e. a stoichiometric mixture of D and L enantiomers of alanine, leucine, lysine, etc., whose spontaneous synthesis from simple precursors like methane, water, carbon dioxide, ammonia, nitrogen would be easily envisaged through mechanisms demonstrated by famous <u>Urey-Miller</u> (1) experiments from interaction of lightning with chemicals on the early Earth. <u>Accidently</u> a crystal nucleus of the one or the other enantiomer was formed which then triggered the fast precipitation of the whole enantiomer content up to thermodynamic saturation; of course the corresponding other enantiomer would remain unprecipitated in the supersaturated solution. And this accidentally crystallized enantiomer were the precursor for further polymerization reaction which would eventually lead to the first "prebiotic" enzymoide structures, which were capable of reproducing into long chains of biological useful entities. After numerous unsuccesfull attempts one of these reaction sets would give rise to the start of the biological elements making up ultimately our recent biosphere which we have been familiar with on our earth. Kondepudi (2) showed a way of enriching a chirally pure biosphere on mere stochastic processes. As a result we arrive necessarily at a world which is accidentely made up of the L-aminoacids and D-sugars. Principally according to this model, the chance of having arrived on the exact mirror image of the recent terrestial biosphere had been precisely identical, random choice had triggered the biosphere as it appeared to us ! The general idea leading to this model has been briefly outlined already by M. Calvin (3) in his book "Chemical Evolution" many years ago.

Now, one could agree completely in an opposite way, - and this is what I favour in this paper !

B) The Determinists

The determination model in contrast to the Random Accident is based on different intuitions:
At first, it is academically more "sympathic" to believe in a strictly logical world: some laws of nature (hidden they may be to us) have ruled our world laéading to the chiral biosphere of today ! And we then can postulate a number of such laws, which result from theory and scarce experimental evidence accumulated up to now:

1) By far the strongest argument so far supported by experimental and theoretical experience comes from the observation that Weak Interaction is Parity Violated. Weak Interaction - one of the four fundamental interactions, namely Strong Interaction, Electromagnetic Interaction, Gravity, and Weak Interaction - manifests itself in the emission of β-particles from radioactive atoms. The β's emitted are spinning around the axis of translation only in a preferred (left) rotation; the mirror image cannot be observed in our matter world. In an "anti-matter world" the opposite would be expected (Lee, Yang, and Wu)(4) which is the logical result of universal symmetry conservation. (On this universal scale symmetry law would be re-established). If we assume an interaction of weak and electromagnetic forces determining the properties of atoms and molecules it would appear attractive to deduce that chiral biosphere is just another macroscopic manifestation of the parity violation in particle physics.

Several workers in the field (beginning from Rein to Sandars, Tranter, Mason, McDermott, and others (5,6,7,8,9)) have calculated the contribution of Weak Interaction to energy content of enantiomer molecules on theoretically ab-initio quantum mechanical calculations, ample material has been published, which is quite disturbing to conventional chemical thinking: The enantiomers of any simple molecules are distinguishable (as a result of Weak Interaction Parity

Violation) by their energy contents, however small these actual differences may be, - e.g. D- and L-alanine show differences of the order of 10^{-17} eV in their energy of formation, etc. Surprising again is the qualitative agreement, that in general the "natural" enantiomers like L-amino acids and D-ribose are the more stable ones with consistently lower energy of formation than their enantiomer counterparts. A considerable number of papers has been published now as to the question whether such extremely small differences would suffice to give the evolution of biosphere such a bias so that it would result ultimately in pure homochirallity of the kind observed. Recently Cairns-Smith (10) has shown that indeed the smallest bias would induce the chiral effect desired given sufficient time and populations numbers. His argument states that the selective pressure for one enantiomer would play the prominente role on the level of already functioning biological organisms (very complex machines relying on one standard norm for economic reasons) and not, as thought before at the very early stage of "prebiotic evolution", the crude assembling process of simple small molecules into more complex but still abiotic organic large molecules of higher negentropy.

2) Another rather well founded idea is based on the assumption that parity violated helically spinning β-particles from radioactive nuclei would interact with prochiral or symmetric molecular mixtures in such a way that the one enantiomer would be radiolysed more readily than the other one: Left-handed fast electrones would fit better into a molecule with left-handed arrangement of atoms around the asymmetric center or plane than with the corresponding right-handed one. A better fit would result in larger absorption coefficients. Of course, this model again has its origin in the characteristic properties of Weak Interaction; only the postulated effect in creating chirality through stereoselective irradiation were a rather indirect interaction. This latter idea goes back to Vester and Ulbricht (11), who proposed this hypothesis shortly after discovery of parity violation in Weak Interaction and undertook several attempts to test the idea through many experiments under various conditions. Several authors thereafter took up similar ideas and investigated irradiation with β-particles and stereochemical inductions, to the greatest part with little success: Either the observed effects were only close to significance and shaded by the extent of natural fluctuations or experimental variations or the results could not be confirmed by independent laboratories.

3) Since long it has been recognized that (circular) polarized light had the potential to induce chirality by interaction with racemic or prochiral compounds. R. Kuhn & coworkers (12) were the first who investigated this line of thought and published many data on significant enrichments of optically active compounds through stereoselective photochemical synthesis or decomposition of racemic and prochiral molecules. The sign of the optical activity achieved depend clearly on the sign of the polarization of the interacting light beam. Since the first articles appeared in the late 30's this mechanism has often been applied to different chemical systems and the earlier data had been

qualitatively confirmed. Although the photochemical reaction with polarized light represents one of the few outstanding cases of an "absolute steric synthesis" (= appearance of optical activity from abiotic sources without the aid of strictly defined biotic material) it serves not as a model for the spontaneous appearance of chirality in the biosphere: There is not an obvious source of unidirectional polarized light on Earth. Indeed, there is polarized light on Earth: Sunlight impounding on oceans' surface is partially reflected, through reflection the light will be partially polarized. In combination with Earth' magnetic field it may be preferentially polarized: left- or right-circular polarized, depending on the direction of the magnetic field and the angle of inflection. But the effect on the North- and South Hemisphere is opposite to each other (Byk, Mörtberg) (13,14), so there is no apparent permanent source of right- or left-circular polarized light. We are left again with the problem how to explain the strongly biased biosphere.

4) A few authors have claimed Gravitation had played the essential role of stereoselective induction of proto-biosphere. But the theoretical basis for this model is weak, the very few experiments along this line (Dougherty) (15) are little convincing.

5) The same holds true for a possible influence of Earth' Rotation on stereoselective synthesis or decomposition. So far there has been very scarce evidence on a sound experimental basis demonstrating rotational stress on stereoselective chemical reactions (Kovacs) (16). Coriolis forces and chemistry, although attractive, seems far-fetched and, if ever demonstrated by independent several laboratories, the direction of influence would again be different on the Northern and Southern Hemisphere.

6) How about a chiral reaction proceeding within a electric field vector - an unidirectional electric field has always been there on Earth: lightning is nothing else than a short cut of the tremendous tension built up between the atmosphere (positive pole) and the surface of Earth (negative pole)? The few attempts to let a chemical reaction go in the field put up between two electrodes have failed to demonstrate any stereoselective effect.

7) There is another interesting suggestion "floating" again and again in the physico-chemical literature with a number of ups and downs in popularity within the scientific community: Earth' magnetic field . We do not treat here the effects of magnetic field on chemical reactions in general - there is a bulk of data on magnetochemistry and magneto-catalytic effects, etc. We mean in this context the idea of stereoselective influence on chemical reactions - synthesis or decomposition. About the theoretical basis of such a model explaining laboratory effects and subsequently the appearance of chirality on Earth there has been a lot of confusion. It is the merit of L. Barron and coworkers (17) to have cleared the confusion to a large extent: He could show that magnetic field alone is not a "true chiral" factor to be considered, but the combination of a natural light beam parallel to a direct magnetic field could well serve as a "true chiral" inductor.

It is conceded however that a magnetic field alone (also not "truly chiral") defined as a pseudo chiral could cause a stereoselective effect, if operating far from equilibrium; let us say as a kinetic effect it would suffice as a chiral agent as long as equilibrium is not reached. (In a certain sense such as the Raleigh distillation process would allow the residue to be in enriched in one way or the other). So in spite of general skepticism raised against the possible influence of magnetic field on stereochemical synthesis or decomposition it appears worth designing some experiments along this line of philosophy.

IS THERE ANY SOUND EXPERIMENTAL EVIDENCE FOR DETERMINISTIC MODEL EXPLAINING THE APPEARANCE OF CHIRALITY ON EARTH ?

I want to list in this chapter just briefly some experimental evidence performed in laboratory which demonstrate positive data as to the absolute synthesis of chiral compounds, which may serve as a support of the above described Deterministic model for explaining the creation of a chiral biosphere on Earth:

Garay (18) showed a selective decomposition of racemic amino acid through irradiation with β-particles from Sr-90 source. Darge, Thiemann and Laczko (19) confirmed this with P-32 as a radioactive source. Merwitz, Akaboshi and Tokay (20,21,22) succeeded in stereoselective radiolysis of solid racemic amino acids by β-particles. Bonner (23) published data on the stereoselective radiolysis of racemic amino acids by both left- and right-polarized accelerated electrons in a asymmetric way. Thiemann and coworkers (24,25) showed stereoselective polycondensation and crystallization of racemic amino acids, which they linked to intrinsic energy differences of enantiomer compounds. Keszthelyi and coworkers (26) investigated selective annihilating of positrons depending on the molar ratios of enantiomers in their substrate target. Kovacs (27) showed the dependence of stereoselective crystallization of $NaNH_4$ Tartrate on the addition of β-emitting sources and stereoselective polycondensation depending on the sence of rotating of the stirrer employed in the process. Darge and Thiemann (28) showed the rotational strength of L-phenylalanine depending on the isotope content. Recently Tokay and Merwitz (29) published data on the significant difference of $^{12}C/^{13}C$ ratios in enantiomer amino acids. Gericke and Dougherty (30,31) were the only researchers which claimed to see an effect of magnetic field on the yield of enantiomers in chemical reactions. Moreover, Dougherty (30) claimed to see a gravitation effect on stereoselective synthesis.

In a recent experimental series undertaken in our laboratory we undertook efforts in seeing an influence of magnetic field through a magneto-chemical catalyst. The idea based on the assumption that theoretically a magnetic field of the order of some Tesla would not suffice to induce a stereoselective effect, which would be large enough to be observed directly through chemical or physical analysis; on the other hand, if an effect was induced, the impact on larger molecules would be more pronounced than on small molecules. So we implied that magnetic effects on polymer structures like certain liquid crystals are well known: They are susceptible to changes in magnetic field and undergo transitions to different phases similar to those induced by pressure, temperature, or solvent. As a liquid crystal we investigated both Tween 80 and Triton X 100, surfactants consisting of polyethyleneoxide derivatives and classified as lyotropic crystals owing their charasteristic to the fact that they change structure with the variation of water dissolved. Within the crystal we imbedded the aryl styrenes which yield upon UV-irradiation the chiral hexahelicenes, or a racemic mixture of chromatooxalates (32). After starting the reaction within a homogenic magnetic field of 1,1-Tesla we did not succeed in seeing any stereoselective effect beyond a statistical error. Yet we saw a significant influence of homogene magnetic field on overall chemical yields and on the pathway of the chemical reaction: In other words, magnetic field changes the transition of cis- to trans-compounds considerably but this was not exactly what we looked for. So even with this idea in mind we failed to confirm any direct influence of magnetic field on stereoselective synthesis (or isomerization in the case of chromatooxalate complexes).

An interesting example of the magnetic field effect was published recently by Takahashi (33): He subjected a racemic mixture of phenylglyoxylic acid to electrolytic reduction with a magnetic field and analyzed an up to 23% enrichment of the one enantiomer product (namely phenylalanine); surprisingly the yield did not depend on the sign of the applied magnetic field. Bonner (34) however tried in vain to reproduce this experiment: Under extremely carefully controlled conditions and even with the strongest magnetic field available (superconducting coil of some 70 Tesla) he could not analyse the lightest enrichment of one enantiomer over the other !

I did not mention here the numerous more or "less successful" attempts to prove non-reproducibility of the other above cited positive experiments. So alas much has to be done yet to convince the scientific community of the theoretical base of Deterministic Model by supporting uniequivocally accepted experimental evidence !

REFERENCES

(1) Miller, S.L., Science 117 (1953) 528.
(2) Kondepudi, D.K., Biosystems 20 (1987) 75.
(3) Calvin, M., Chemical Evolution, Oxford Univ. Press, Oxford 1969.
(4) a) Lee, T.D., Yang, C.N., Phys. Rev. 104 (1956) 254.
b) Wu, C.S. et al., Phys. Rev. 105 (1957) 1413.

(5) Rein, D., Some Remarks on Parity Violation Effects of Intramolecular Interactions in Generation and Amplification of Asymmetry in Chemical Systems, (ed. W. Thiemann) Jülich Conference - 13,1974, 91.
(6) Hegstrom, R.A., Rein, D.W., Sandars, P.G.H., J. Chem. Phys. 73 (1980) 2329.
(7) Tranter, G.E., Nature 318 (1985) 172.
(8) Mason, S.F., Tranter, G.E., J. Chem. Soc. Comm. (1983) 112.
(9) McDermott, A.J., Tranter, G.E., Croatica Chem. Acta 62/2A (1989) 165.
(10) Cairns-Smith, A.G., Chemistry in Britain 22/b (1986) 559.
(11) Vester, F., Ulbricht, T.L.V., Kranchi, H., Naturwiss. 46 (1959) 68.
(12) Kuhn, W., Knopp, E., Naturwiss. 18 (1930) 183.
(13) Byk, A., Z. Phys. Chem. 49 (1904) 641.
(14) Mõrtberg, L., Nature 232 (1971) 105.
(15) Dougherty, R.C., Origins of Life 11 (1981) 71.
(16) Kovacs, U.K., Keszthelyi, L., Goldanskii, V.J., Origins of Life 11 (1981) 93.
(17) Barron, L.D., Chem. Physics Letters 135 (1987) 1.
(18) Garay, A., Nature 220 (1968) 368.
(19) Darge, W., Laczkó, J., Thiemann, W., Nature 261 (1976) 522.
(20) Merwitz, O., Rad. Env. Biophysics 13 (1976) 63.
(21) Akaboshi, R., Noda, R., Kawai, U., Maki, H., Ito, V., Origins of Life 11 (1981) 23.
(22) a) Tokay, R.K. Norden, B., Liljenzin, J.O., Andersson, S.J., J. Radioanalyt. Nucl. Chem. 1986
b) idem. Council Europe, Newsletter 12 (1986) 8.
(23) Bonner, W.A., Van Dort, M.A., Yearian, M.R., Nature 258 (1975) 419.
(24) Thiemann, W., Darge, W., Origins of Life 5 (1974) 263.
(25) Thiemann, W., J. Mol. Evolution 4/1 (1974) 85.
(26) Keszthelyi, L., Origins of Life 11 (1981) 9.
(27) Kovacs, K.L., ibidem. 11 (1981) 37.
(28) Darge, W., Lasczkó, J., Thiemann, W., J. Radioanalyt. Chem. 30 (1976) 521.
(29) Tokay, R.K., Merwitz, O., Priv. Comm. to be published.
(30) Dougherty, R.C., Origins of Life 11 (1981) 71.
(31) Gericke, P., Naturwiss. 62 (1975) 38.
(32) Teutsch, H., Zur Entstehung ostischer Aktivität in der Biosphäre, Ausgewählte Photoreaktionen als Asymmetrische Synthese, Dissertation Univ. Bremen , Bremen 1988. To be published in Origins of Life (1990).
(33) Takahashi, F., Tomii, K., Takahashi, H., Electrochim. Acta 31 (1986) 127.
(34) Bonner, W.A., Electrochim. Acta (1989) in press.

ORIGIN OF TRACE ELEMENT REQUIREMENTS BY LIVING MATTER

Vlado Valkovic

Rudjer Boskovic Institute

Bijenicka 54, 41000 Zagreb, Croatia, Yugoslavia

1. INTRODUCTION

Biological systems are the systems that use energy flow through them to create order. In biological processes each step is carefully marked in terms of free energy before and afterwards. On thermodynamics term the change of entropy is minimal. Evolution is driven by minimizing the time rate of an increase in entropy (ds/dt --> minimal).

On the other hand, biological systems take up elements from the environment in which they live in. This is indicated schematically in Fig. 1. There are two types of elementary cycles: organic and inorganic, depending on whether or not an element makes and breaks covalent bonds with carbon during its cycle. Living matter uses the energy of sunlight to carry the chemical elements in the environment through continuous cycles of activity.

The organic cycle includes also the eleven elements: H, C, N, O, Na, Mg, P, S, Cl, K and Ca which form the bulk of the living matter. They all have very low atomic weights and they belong to the lowest 20 elements of the period table. These elements are also the most abundant elements in the Universe. The only nonessential light elements, with exclusion of noble gases, are Li, Be, B (boron is essential for some plants) and maybe Al.

In addition, a number of elements have been recognized as essential trace elements. For example, essential trace elements for warmblooded animals are: F, Si, V, Cr, Mn, Fe, Co, Ni, Cu, Zn, Se, Mo, Sn and I. For the most part, the essential trace elements are transition metals with unfilled d-orbitals. None of the 39 elements beyond iodine (Z=53) has ever been shown to have any physiological significance.

Symmetries in Science IV
Edited by B. Gruber and J. H. Yopp
Plenum Press, New York, 1990

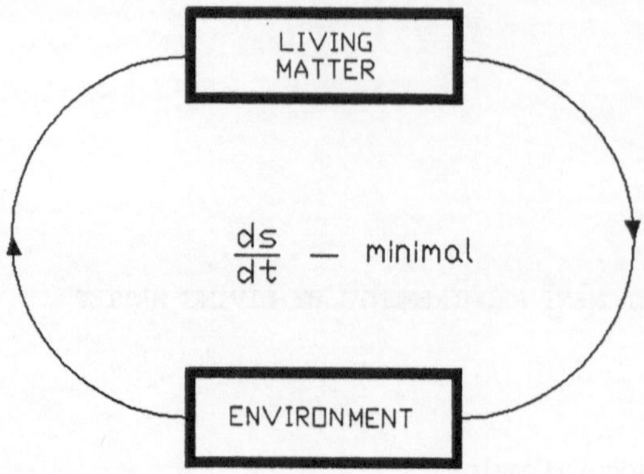

Fig. 1. Elementary cycles also have to satisfy the requirement on time rate of the change of entropy of a system to be minimal.

It is impossible to replace essential trace elements with something else in the living matter. This can be stretched to the point that the maximum number of living organisms (in a community) could be evaluated by the availability of some essential trace elements from the environment.

The distribution of essential elements within the periodic table of elements might contain some clues about the origin of element requirements and possibly about the development of life in prebiotic time. The assumption that the vital body chemistry should bear similarity to the primordial chemical environment is a reasonable one. An organism would not make itself dependent on a rare element for its existence providing a more abundant element could play the same role.

It is obvious that the presence of an element is a necessary prerequisite for the development of an essential metabolism based on that element. In explaining why a particular element has been selected for an essential biochemical role, one of the possible approaches is as follows: If an element has not be selected, this may be because its abundance in the available environment is too low, either on the absolute scale, or in comparison with some other element that can play the same role.

In our discussion we shall assume that the distribution of essential elements reflects conditions of the primordial chemical environment.

Dependences of concentration factors for essential trace elements on some physical parameters are indicators of conditions under which the choice of essentiality has been made.

2. CONCENTRATION FACTOR

One of the most important characteristics of the living matter is its ability to take up elements from the environment against the concentration gradient (Valković, 1980). Assuming that a concentration of an element i in the environment is C^i_{ENV}, in the living organism C^i_L, then the concentration factor can be defined by

$$C^i_L = F^i(x,y,z,...)\, C^i_{ENV} \tag{1}$$

The organisms concentrate all elements present in their environment (see Fig. 2). However, the uptake of the elements from the environment depends

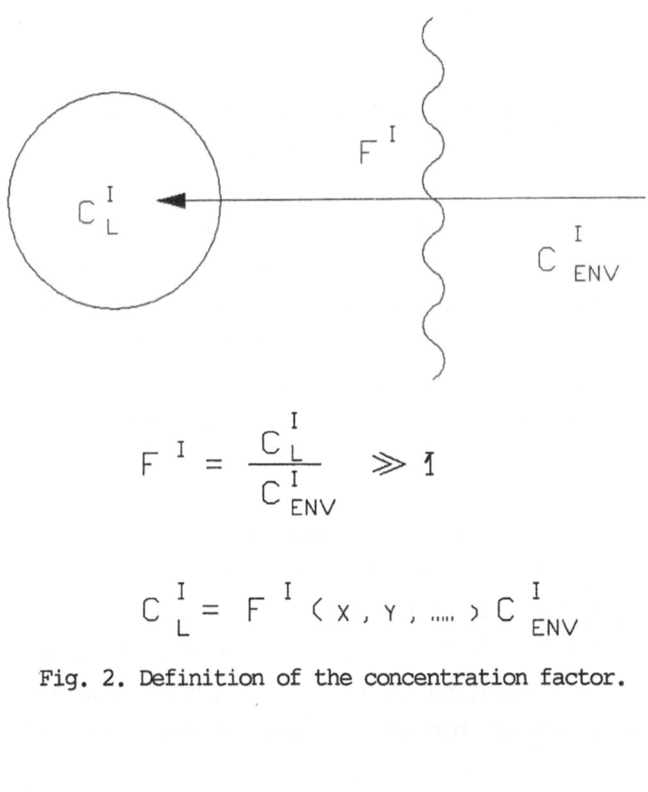

Fig. 2. Definition of the concentration factor.

upon many factors. For example, the plant uptake from the soil is determined by: the element abundance in the lithosphere, the form of element, the pH of the soil, the physical conditions of soil (tilth, temperature, moisture content) and the genetic constitution of the plant.

Higher animals and man have developed homeostatic mechanisms (in gut and kidney) in order to conserve deficiencies and reject excesses. Only a portion of the nutrients which are ingested are absorbed and utilized in the metabolism of individual humans and animals. Many factors within the digestive tract may influence the absorption of any element, and the interactions may be very complex. A very important factor is the relative concentration of the different elements with respect to one another. It seems that different ions may follow the same metabolic pathway across the intestinal membranes.

It should be emphasized that a change of the concentration of an element in the organisms, C_L^i, could result either from the change in the concentration of this element in the environment, C_{ENV}^i, or from the change of values for the parameters x,y,z,... on which the concentration factor depends:

$$\Delta C_L^i = F^i \Delta C_{ENV}^i + C_{ENV}^i \left[\frac{\partial F}{\partial x} \Delta x + \frac{\partial F}{\partial y} \Delta y \ldots \right] \qquad (2)$$

Concentration factors are most easily determined for marine organisms. For example, Fisher (1986) has reported volume/volume concentration factors (mol metal μm^{-3} cell/mol metal dissolved μm^{-3} ambient seawater, at time of equilibrium) for metals in several species of marine phytoplankton, see Fig. 3. Concentration factors vary among metals from $\simeq 0$ to 10^6, and there are only modest differences (less than an order of magnitude) among algal species for any particular metal. Regression analyses showed that, at equilibrium, the logs of the concentration factors are exponentially related to the solubility products of metal hydroxides. It seems that F^i in phytoplankton can be predicted to within about half an order of magnitude on the basis of the metal affinity for hydroxyl groups.

Laboratory experiments (Fisher and Fowler, 1987) have shown that marine plankton greatly concentrate even transuranic elements (Pu, Am, Cm, Cf) up to 10^6 times in phytoplankton, 10^2-10^4 times in crustacean zooplankton, and 10^1-10^2 times in appendicularians. It was found that correlations of concentration factors for transuranic elements with surface-volume ratios for organisms were described by the same equation

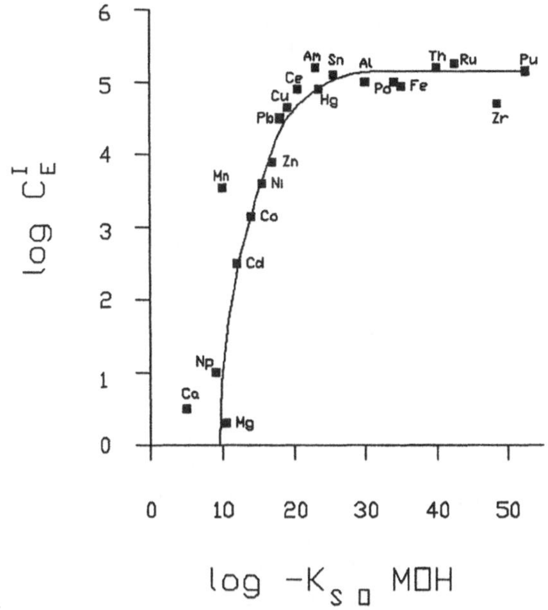

Fig. 3. Average correlation of C^i_E of metals in marine phytoplankton with the logs of the solubility products ($-K_{SO}$) of metal hydroxides (after Fisher, 1986).

for phytoplankton, euphausiids and copepods, suggesting that the uptake process in these organisms proceeds by passive adsorption onto surfaces, and that the surfaces of these organisms have approximately equal affinities for these elements (see Fig. 4.).

3. ESSENTIAL TRACE ELEMENTS

It seems that the organisms concentrate all elements present in the environment. With improved experimental techniques probably all the elements could be found in the tissues and liquids within a living organism. The question is, "How many of them are essential for life?" Obviously, some are acquired only as environmental contaminants and reflect the contact of the organism with its environment.

The bulk of living matter consists of eleven elements which have very low atomic weight: H, C, N, O, Na, Mg, P, S, Cl, K, and Ca. These elements have been known to be essential for life for a long time because their presence is easy to detect. Others occur in such small amounts in

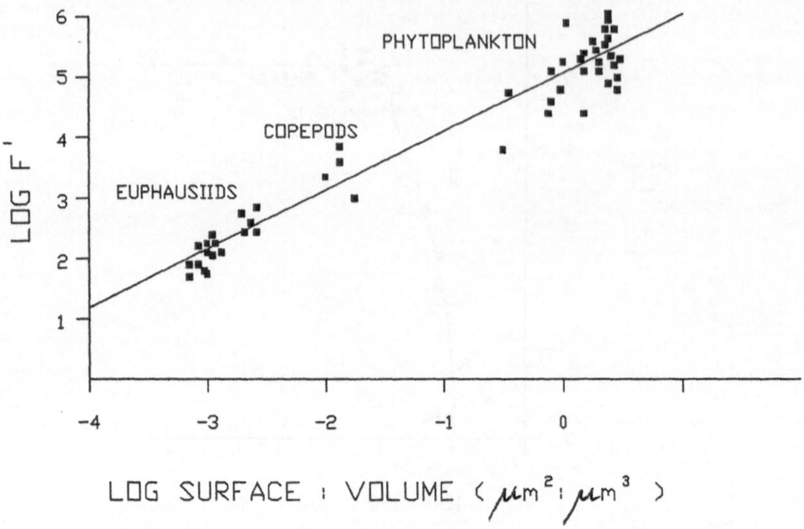

Fig. 4. Correlation of log F^i for transuranic elements (Am, Pu, Cm, Cf) with log surface/volume ratio for marine plankton (after Fisher and Fowler, 1987).

living tissues that for a long time the precise concentrations were not measured and they were often mentioned as occurring in "traces".

In order for an element to be essential for life it must satisfy some criteria. According to Cotzias and Foradori (1969) they are as follows:

1. It is present in healthy tissues of all living things.
2. Its concentration from one animal to the next is fairly constant.
3. Its withdrawal from the body induces reproducibly the same structural and physiological abnormalities regardless of the species studies.
4. Its addition either prevents or reverses their abnormalities.
5. The abnormalities induced by deficiency are always accompanied by pertinent biochemical changes.
6. These biochemical changes can be prevented or cured when the deficiency is prevented or remedied.

The list of essential trace elements has been established from the studies of trace elements requirements of plants and animals. These studies have been performed in large numbers by withdrawing or adding some of the elements from the diet in an otherwise controlled environment. Progress in this field is closely tied to the modern developments in trace element analysis and the introduction of ultraclean isolator techniques.

The relation of the uptake of essential elements to growth may be presented as in Fig. 5a. This functional relationship between the concentration of an element and its effects may be considered as a definition of essentiality. There is a rather narrow range of adequacy of element in the organisms. Smaller concentrations result in different abnormalities induced by deficiencies which are accompanied by pertinent specific biochemical changes. Higher concentrations result in toxicity. It is obvious that all essential elements may have toxic effects if present in high enough concentrations. Elements without known biological functions will not have adequate range of concentration. In this case one may speak about toxic range only, and the curve in Fig. 5b has only a part corresponding to the decreasing yield or growth with the increase in element concentration.

In plants, it is possible under severe deficiency conditions, for a decrease in the concentration of an element to result in a small increase in growth (Steenbjerg, 1951).

Organisms isolated from environments polluted by heavy metals are often tolerant of those metals. Tolerance is usually accompanied by metal uptake equal to or greater than that of nontolerant organisms; the accumulated metals seem to be chemically detoxified and/or physically sequestered to render them inactive. Because heavy metal tolerant algae plants, yeast, and invertebrates have been found not to restrict metal uptake, metal exclusion has been considered a rare mechanism of tolerance.

The present knowledge on essential bulk and trace elements is summarized in Fig. 6. The list of elements recognized as essential for at least some forms of living matter include: (i) bulk elements H, C, N, O, Na, Mg, P, S, Cl, K, Ca, and (ii) trace elements B, F, Al (?), Si, V, Cr, Mn, Fe, Co, Ni, Cu, Zn, As, Se, Br (?), Mo, Sn and I. Only three of the 24 elements essential for life have an atomic number above 34. They are essential trace elements: Mo (Z = 42), Sn (Z = 50), and I (Z = 52).

A great majority of the trace elements serve chiefly as key components of enzyme systems of proteins with vital functions. Enzymes in which metals are tightly incorporated are called metalloenzymes, since the metal is usually embedded deep inside the structure of the protein. If the metal atom is removed, the protein usually loses its capacity to function as an

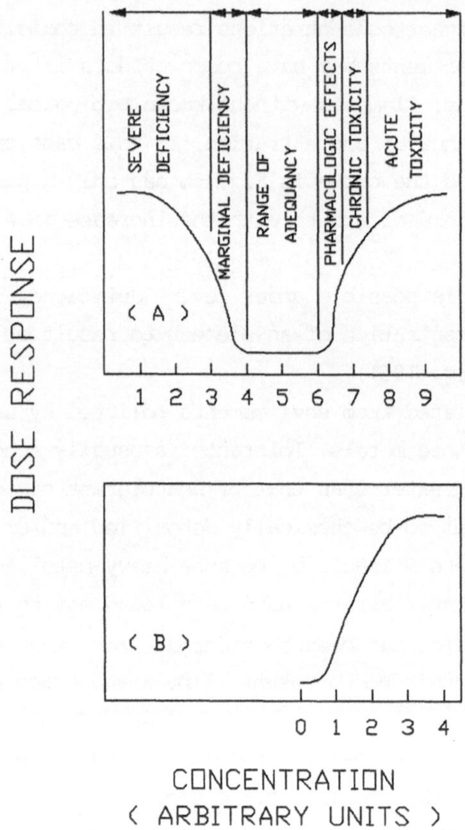

Fig. 5. Dose response curve for (a) an essential element and (b) a toxic element.

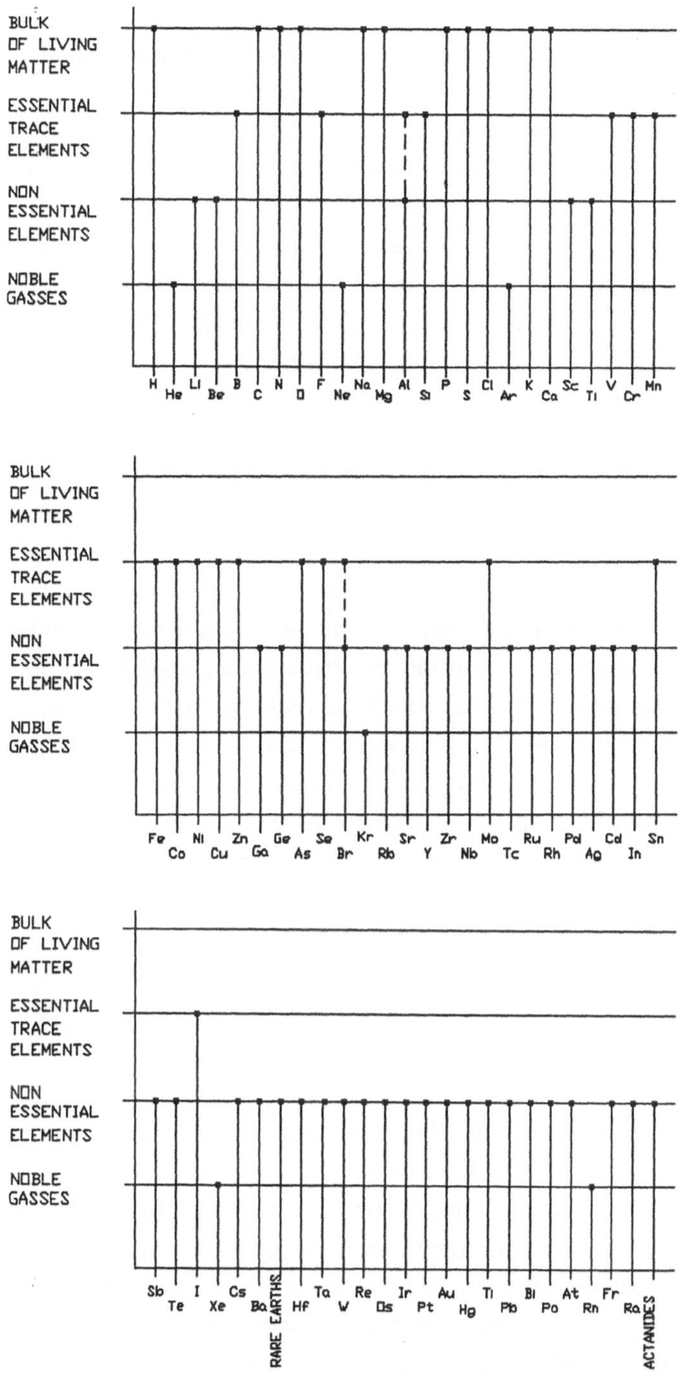

Fig. 6. Present knowledge about elements essential for living matter.

enzyme. As pointed out by several research groups, many clinical and pathological disorders arise in the animal as a consequence of trace element deficiencies and excesses for which there are no acceptable explanations in biochemical or enzymic terms. This suggests either that there are many trace element-dependent enzymes of great metabolic significance which are not yet discovered or that these elements participate in the activity of other compounds in tissues.

Very intriguing are multiple and complex interdependencies of elements. This might suggest that there are some not yet discovered important facts about the role and interrelations of trace elements.

There are many examples indicating clearly that metals are able to exercise control over catalytic reactions both on a quantitative and qualitative basis. Actually, a metal can influence the activity of an enzyme by its sheer physical presence in the active center of the enzyme. It is not unlikely, that even when a metal is attached to an enzyme at a distance from its active center, configurational alterations may affect the activity of the enzyme (Cotzias and Foradori, 1969).

Wacker and Vallee (1959) were the first to show convincingly the presence of metals in nucleic acids macromolecules. They found chromium, nickel, and manganese in ribonucleic acid (RNA) preparations derived from a wide variety of sources. It is important that the total concentration of the metals bore a constant ratio to phosphate (1:50 for RNA and 1:150 for DNA). If these metals exercise control over the configuration of nucleic acids by stabilizing the secondary helical structure in a manner similar to disulfide bridges, then they must be considered to play a role in protein synthesis and in the transmission of genetic information.

4. PORPHYRINS

As an example of biologically important molecules whose central part is an atom of metal we shall mention porphyrins. The uniqueness of porphyrins in living matter arises from their importance to many life processes such as photosynthesis (chlorophyll) and respiration (cytochrome). Porphyrins are heterocyclic aromatic molecules, of almost planar geometry and high symmetry. The center of a molecule contains a divalent metal atom; the removal of the center atom and its replacement by two hydrogen atoms cause it to become a "free base" porphyrin.

The metal of chlorophyll is magnesium. The pigment of chlorophylls possesses an absorption spectrum in the visible region, and enables these radiations to be utilized for the photosynthesis by the reaction:

$$h\nu + 6\ CO_2 + 6\ H_2O \longrightarrow C_6(H_2O)_6 + 6\ O_2 \qquad (3)$$

There are various types of chlorophylls. All of them can be characterized by absorption spectra having two maxima, and a minimum in the region of 500 - 650 millimicrons wavelength. This is an intriguing fact because it corresponds to the maximum in the energy distribution of the sunlight at the Earth's surface. The two important types of chlorophylls have the following formulae: $C_{55}O_5H_{72}N_4Mg$ and $C_{55}O_6H_{70}N_4Mg$.

The appearance of chlorophylls on the planet Earth has marked a decisive stage in the evolution of the living world. Large amounts of oxygen were liberated and the Earth's atmosphere was created. Since then life has been protected by the ozone and oxygen, and respiration was able to become aerial.

Porphyrins were among the first large molecules identified in the interstellar space. A specific porphyrin molecule identified is $C_{46}H_{30}Mg\ N_6$ (bispyridylmagnesiumtetrabenzporphine - called χ, (Johnson, 1972). It was postulated by the author that its synthesis and destruction proceed in dense gaseous lacalized regions in space, near an intense infrared radiation source. This molecule has an enormous thermodynamic stability, binding energy of $\sim$ 17 eV (compared with the dissociation energy of water of $\sim$ 10 eV), and can therefore succeed in a battle of "molecular survival".

5. ELEMENT SYNTHESIS

The origin of the matter is the Universe and the origin of the chemical elements and their abundances are two related subjects. Our present knowledge of the Universe is based on the observation that the distant galaxies are moving away from our Galaxy with a velocity proportional to the distance. We live in an expanding Universe. There are several types of cosmological theories speculating about this. The theories of a single sudden origin in a "Big-Bang" are the least controversial. Since the discovery of the isotopic microwave background radiation, they have been generally accepted. The element synthesis is believed to take place in the stars, which at the beginning were made of hydrogen (^{1}H) only.

Everything starts with hydrogen burning (or the proton-proton chain) which is a sequence of nuclear reactions with the final result

$$4\ ^{1}H \longrightarrow\ ^{4}He + \text{energy} \ . \qquad (4)$$

About 4.2×10^{-12} J of energy is realized any time four hydrogen atoms are converted into a helium atom. Stars spend most of their energy in this phase during which they convert hydrogen into helium in hot central regions.

In proton-proton chain no ^{6}Li, ^{7}Li, ^{8}Be, ^{10}B isotopes are formed. It is believed that our Sun is in the stage of burning up its hydrogen. However, when a star of the same mass as our Sun spends about 10% of its hydrogen in producing ^{4}He, thermal pressure is not great enough and gravity becomes the dominant factor. As a consequence gravitational contraction takes place and the radius of the star decreases. As a result of gravitational contraction the density of the star will increase. This results in a temperature increase and a consequent increase in the velocities of particles. This is the beginning of the formation of heavier elements. The process is "He-burning". In the sequence of nuclear reactions a variety of heavier elements and isotopes up to Ti are formed, including C, O, Ne, Mg, Si, S, Ar and some others.

A star cannot remain permanently stable since it loses the energy by the emission of electromagnetic radiation and by emission of particles. Finally, gravitational contraction will overcome all other forces. After 10^{6} to 10^{10} years of slow contraction, the star must die. The most common way is by a nuclear explosion which disintegrates the object. The natural abundances of most elements are probably established in the few final seconds of a star's lifetime as it disrupts explosively. Nuclear reactions burn so furiously at the high temperatures of the explosion that the composition of an entire star is altered. All the matter is thrown out into the interstellar space in the form of dust particles. Somewhere, sometimes, the residues of many such explosions will form a cloud of gas which will be the nucleus for the formation of a new star. According to present theories the star can also die by a gravitational collapse.

In the sequence of these processes the amount of heavier elements will increase relative to lighter elements. Taking into account a closed system, such as galaxy, one can speak about time variations of universal abundance curve of elements.

Spectroscopic analyses of the stars' atmospheres reveal that, at the surface of most stars, the distribution of elements heavier than helium is, to a first approximation, independent of the total abundance of the heavy elements, even though this later quantity varies by three orders of magnitude as one ranges from the most metal rich stars in the galactic disk to the most metal poor stars in the outer fringes of the halo.

If one assumes that the element formation occurs at singularities (novae and supernovae), then the existence of an approximately universal distribution for the heavy elements implies one of two things: either (1) the distribution of elements emerging from each singularity is the same, or (2) products expelled from a multitude of singularities mix at a rate that is fast relative to the rate at which new singularities are born out of the steadily diminishing supply of interstellar gas. It seems that the second alternative is the more likely one.

The total abundance of heavy elements at a star's surface seems to be roughly correlated with the star's age and with its mean (rms) distance above the galactic plane. In general, the older the star, or the greater its mean distance above the plane, the lower is its metal abundance. There also appears to be a distinct difference between stars in the halo and stars in the disk of the galaxy. Most of the stars in the halo are metal poor and are, within a spread of perhaps a billion years or so, all about 10^{10} years old. Most of the stars in the disk are comparable to the Sun in metal richness and have ages that run the entire range from 10^{10} years to 0 years.

6. ABUNDANCES OF ELEMENTS

In addition to the Earth, the Moon and the meteorites, the abundances of at least some elements can be obtained for the solar system, for the Sun and other stars, for gaseous nebulae, including some in external galaxies, and for interstellar space and cosmic rays. For all these parts of the Universe for which the abundances of the elements can be found, hydrogen (isotopes ^{1}H and ^{2}H) and helium (isotopes ^{3}He and ^{4}He) are much more abundant than all of the elements put together.

There are serious difficulties in obtaining data on the composition of the Universe. Let us illustrate this with studies of elemental composition of our planet. The most essential shortcomings of the studies of the chemical composition of the Earth's crust are the lack of allowance for the spatial distribution of the rocks and for the nature of the distribution of elements in various rocks. The numerical data of the various authors were frequently obtained under highly diverse conditions and by very different methods. Moreover, in numerous cases these data refer to rocks of regions randomly chosen, whereas extensive territories remain unmapped. It follows that the data used in calculating the average content of chemical elements are essentially heterogeneous due to the different sampling and analytical methods. It should be noted that the average composition of the Earth is determined entirely by that of the matter below the crust. Earth crust abundances do not indicate the primordial composition of the solar system. With respect to this problem, among the stony meteorites, or erolites, the chondrites compromise the most important class. These objects are believed by some to represent the best sample available to us of nonvolatile materials out of which the solar system has been formed. The surface of the sun is believed to represent rather well the primordial composition of the solar system.

The trace element content of meteorites combined with the isotopic composition of the individual elements, showed in a strikingly unambiguous way that the composition of meteorites resembled the products of nuclear reactions that had led to the formation of the elements in their relative proportions as we now find them in nature. There is remarkable agreement in elemental composition of the sun and the meteorites. One exception is iron, which is lower by a factor of 3 to 4 in the sun. Some elements such as Zn, Cd, Sn, In, and halogens show low and irregular values in meteorites.

In an effort to construct the universal abundance curve of the elements one has to include information on the composition of the stars. Although it seems that the composition of most stars resembles that of the sun fairly closely, this can be misleading since these conclusions are made from the observation of light spectra only. Such measurements can give information about the surface only. The interior of stars has probably a different composition. Present universal abundance curve of elements is probably best constructed from a very well known compilation of elemental abundances by Suess and Urey (1956). The Suess-Urey abundance curve is a semiempirical one; it is based partly on the measured abundances and on the consideration of the nuclear stability (see Fig. 7.).

The abundance distribution of light elements is characterized by a deep minimum for Li, Be, and B. The explanation for this is found in the properties of nuclear reactions taking place during the hydrogen burning (p-p chain). During this process no Li, Be, and B nuclei are made. The next nucleus made is ^{12}C (as a result of the He-burning). Another deep minimum occurs for Sc.

The elements from these abundance curve minima are not essential for life, with the exception of boron, which is essential for plants. This fact should not be taken as a pure coincidence. However, the relative abundance of the elements for the cosmic rays near the Earth does not show these minima. It is believed that different spallation processes are responsible for filling in these minima. Although this could be the case with Li-Be-B minimum, such an argument is not nearly acceptable for the nuclei in Sc-region.

The striking property of abundance curve is the iron peak, being five orders of magnitude above the Li-Be-B and Sc minima. This peak corresponds to a maximum nucleon binding energy. From iron on neutron capture processes are responsible for element synthesis, and element abundance curve levels off around 10^{-7} Si abundance, with only small Ba and Pt peaks. In general, the abundance of a particular atomic nucleus depends on its inherent stability and binding energy and also on the effectiveness of the nuclear process by which it is believed to have been formed.

The planet Earth, though 4.5×10^9 years old is not yet in chemical equilibrium. The core is thought to consist mainly of iron and a few percents of nickel with an average density of 10^4 kg m^{-3}. It represents about 31.4% of the mass of the Earth. The mantle may consist of the material similar to the lava from the Hawaiian volcanoes (mainly magnesium silicate, density 5×10^3 kg m^{-3}). The mantle represents about 68.1% of the mass of the Earth. The crust of the Earth extends to a distance of only 0.08% of the depth to the center of the Earth.

The materials which form the igneous rocks crystallize during the cooling process of a liquid magma. As time goes on, igneous rocks are slowly destroyed by physical, chemical, and biological processes. Sedimentary rocks comprise about 5% of the lithosphere, while the igneous rocks form 95%. The three main types of sedimentary rocks are shale, sandstone, and limestone, and their relative abundance determined from geochemical data ranges from 70 to 83% shale, from 8 to 16% sandstone, and from 5 to 14% limestone.

Crustal abundance curve (Parker, 1967) is dominated by noble gases minima and severalpeaks most notably O, Al-Si, K-Ca, Fe, Rb-Sr, Sn, Ba and Pb (see Fig. 8.).

Of special interest is the elemental composition of seawater. The primordial earth may have had altered abundances of some elements in the ocean and atmosphere. It may be that the present concentrations are not unlike those of the primitive earth but the situation is clearly complex.

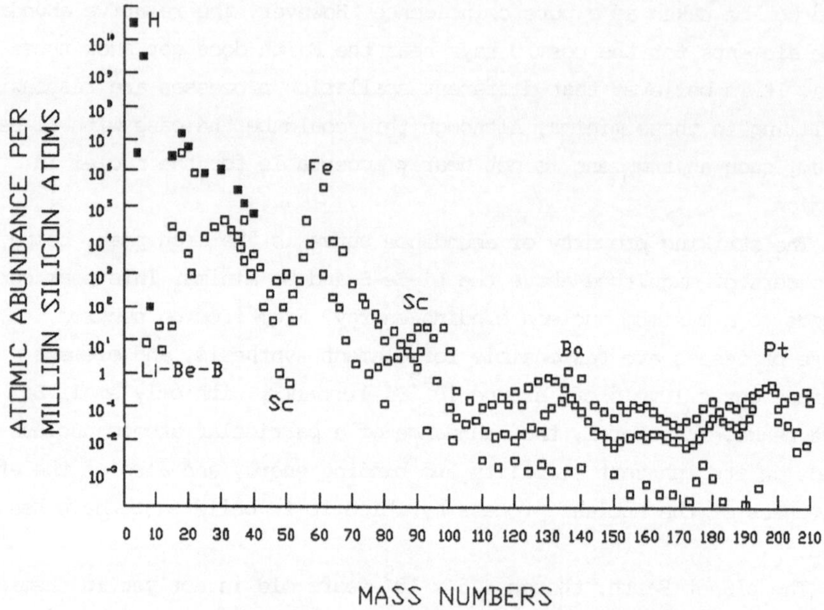

Fig. 7. Universal abundance curve of elements.

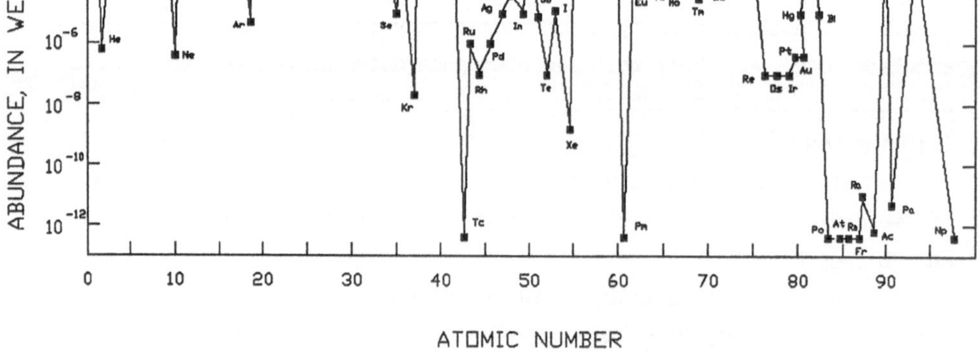

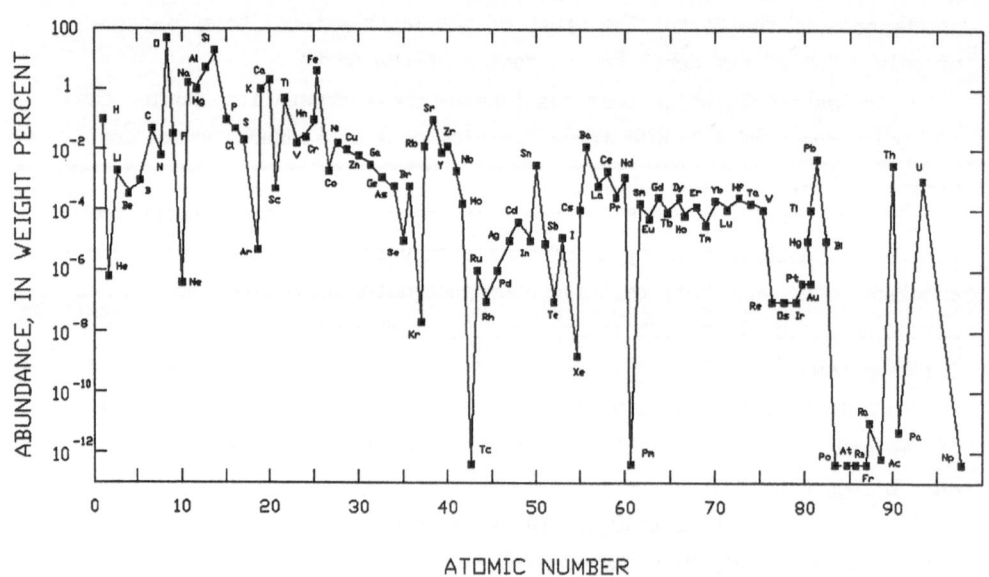

Fig. 8. Crustal abundance curve of elements.

Oceans are modified by many cycles of erosion and sedimentation, and it is thought that these two processes have been in dynamic equilibrium for at least 0.5 x 10^9 years so that the composition of the ocean has remained the same for that length of time.

The existence of the oceans has been estimated to be about 2.7 x 10^9 years. There are two acceptable theories on the origin of oceans. One assumes an early release of water from the crystal rocks, the other assumes the continuous release of water through geological time. Most of the properties and processes of the oceans involve in some way their chemical composition. Large numbers of elements, ranging in concentration through 14 orders of magnitude, are present in oceans. The elemental composition of seawater is presented in graphical form in Fig. 9 (based on data from Bowen, 1966). The only elements heavier than Ca present in concentrations larger than 1 ppm are Br and Sr. The concentration of metals which serve as essential trace elements in living matter is of the order of 10^{-3} ppm. Considerable uncertainty exists on the precise form of trace elements in seawater. It should be mentioned that the concentrations of many elements in the oceans are variable. Some elements show strong regional variations, while some others are fairly constant in concentration.

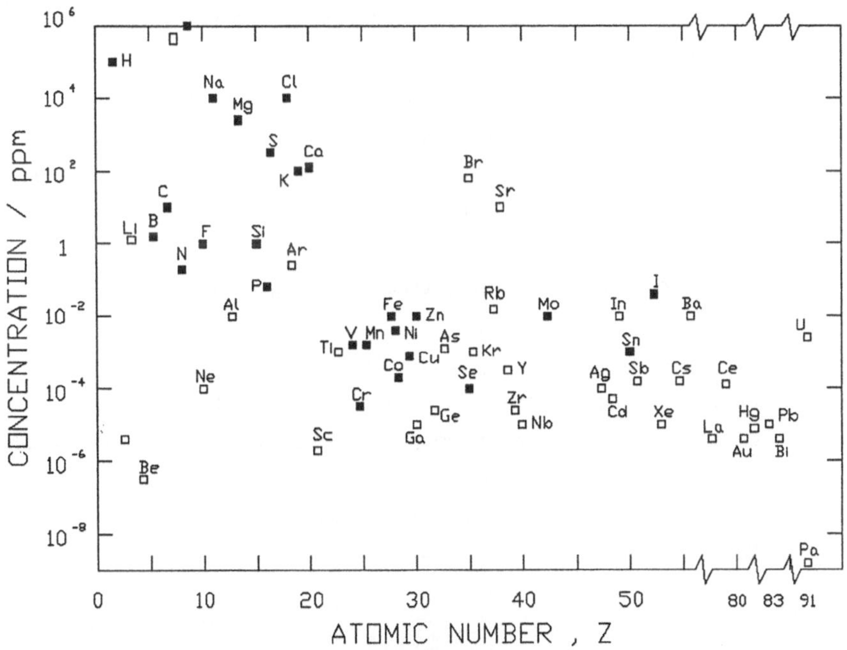

Fig. 9. Elemental composition of seawater.

7. ORIGIN OF ELEMENTAL REQUIREMENTS

The suggestion that the concentrations in seawater were those crucial to the origin of elemental requirements has been often discussed (see Chappell et al., 1974; Jukes, 1974; Orgel, 1974; Banin and Navrot, 1975). The ocean is thought to be in a steady state with sediments and the atmosphere. Although it is often remarked that the composition of organisms resembles that of the ocean, one might get different conclusions from Fig. 9. High concentrations of nonessentials: Li, Br, Sr, Rb, and some others are evident. Metals which serve as essential trace elements are very strongly depleted in the ocean compared to plant or animal requirements. In addition, the content of both carbon and nitrogen is low compared to living matter concentrations and cosmic abundances.

We have already stated that the presence of an element is a necessary prerequisite to the development of essential metabolism based on that element. Two questions should be answered first in such an approach:

(i) Which environment should be compared to the composition of living organisms? The answer: That with which the organism has an intimate contact.

(ii) Which organism should be chosen for such a study? The answer: It seems that the appropriate organism would be the one that is very simple (for example, algae).

Abundance as a factor in the origin of mineral nutrient requirements has also been discussed by McClendon (1976). He has considered elemental concentrations in the environment of the living species. According to his estimates there appears to be an absolute limit in abundance below which no element may fall and still be available in adequate amounts even as a micronutrient (10 - 20 μ moles/kg crust or 1 - 2 nM in the ocean). It was postulated that all of the required elements fall under one of the four hypotheses:

1. Unique requirement dating from the origin of life; H, K, Mg, C, N, O, P, S, Fe.
2. Unique requirement acquired later; B, Se, I.
3. Primordial requirement which was satisfied by a number of elements - evolutionary adoption being made to the most abundant member; K vs Rb, Mg vs Be, S vs Se, Cl vs Br, H vs F.
4. The same as under 3. but a later acquisition; Ca vs Sr, Na vs Li, Si vs Ge.

According to Egami (1974) a good correlation can be found between the biological behaviour of different elements and their concentration in seawater. This suggests the hypothesis that the composition of the present seawater reflects that of the primeval seawater at the time of the evolution of these enzyme systems. A concentration in the seawater of about 1 - 5 nM may be regarded as "critical". Elements with concentrations in seawater above this critical concentration could influence early evolutionary events, and so became either essential or neutral elements. Organisms evolved independently of trace elements with concentrations less than the critical concentration.

The concentration of elements in seawater can be divided into several ranges:

Range 1	Range 2	Range 3	Range 4	Range 5
$C > 10^6$ nM	$10^6 > C > 10^2$ nM	$10^2 > C > 5$ NM	$5 > C > 1$ nM	1 nM $> C$
all bulk elements except P	P, most bioelements for specialized organisms	trace elements participating in enzyme structure	critical concentration range	range of elements unrelated to evolution

Since it is generally believed that the early stages of the evolution of life forms occurred in the sea, it would be reasonable to assume that the concentrations of trace elements in seawater would be more relevant than crustal abundances (Chappell et al., 1974).

In a similar vein, this author suggested that aluminum might be an essential element for life, because the abundance of aluminum in environments suitable for the origin of life is relatively high (Valkovic, 1980). To discuss the essentiality of the elements on the basis of their abundance alone may not reveal the entire story, as we shall see later. Banin and Navrot found that, when elemental enrichment factors in living organisms were plotted against the ionic potential of the elements, the pattern was strikingly similar to that found in present-day seawater. They concluded that this observation supports the hypothesis that life began in a water--rich environment, i.e., in primitive seawater (Banin and Navrot, 1975).

Contrary to all expectations universal (cosmic) abundance curve of elements is in the best agreement with the distribution of essential elements within the periodic table. Essential elements are the most abundant elements; essential trace elements are almost all grouped in the secondary peak (around Fe). This is the region of the maximum nucleon binding energy in the nucleus, the fact which is responsible for the peak in the universal abundance curve. Obviously the abundance of an element is not the only factor determining the biological role. Let us illustrate the effects' solubility in water in the case of porphyrins.

Factors influencing the selection of the appropriate central metal atom in the porphyrin ring in the process of evolution are discussed by several authors, see, for example, Krasnovsky, 1976). The compounds of Si and Ti are practically insoluble in water; this insolubility probably prevented the incorporation of these metals into porphyrin complexes. Complexes of Na, K, and Ca with porphyrins are hydrolyzed in aqueous media. Only the complexes of magnesium and iron are sufficiently stable to be perpetuated in the course of evolution. These complexes have auxilliary coordination vacancies perpendicular to the plane of the porphyrin ring and thus are also capable of binding reacting molecules such as proteins.

Metals of low electronegativity and low oxidation numbers activate the porphyrin ligand with respect to oxidation and deactivate it with respect to reductions. The opposite is true for metals of high electronegativity and/or oxidation number.

Such reversible reactions of the macrocyclic ring are of importance in biological systems containing magnesium porphyrins. On the other hand, the electron-rich transition-metal ions with formally high oxidation numbers are stable in the porphyrin cavity and tend to transfer some electrons to axial ligands. This effect is probably of biological importance in the hemoproteins, which function in electron tranfers and oxygen activation. V and Ni porphyrins have been found in petroleum; they are probably formed as a result of chlorophyll degradation and central Mg atom substitution.

If we accept the idea that the chemical composition of living organisms must reflect to some extent the composition of the environment in which they evolved, the presence in living organisms of elements that are extremely rare on the Earth might indicate that life is extraterrestrial in orgin. Several years ago the question of extraterrestrial origin of life has been reopened, with the abundance of molybden as a factor in the argument (Crick and Orgel, 1973; Orgel, 1974).

These authors considered the possibility that life reached the Earth from other planets. One major reason why they thought this was so is that the concentration of molybdenum in the Earth environment is lower than that of chromium or nickel. They conclude from this that life originated in an environment where molybdenum was likely more abundant than chromium or nickel. Certainly, molybdenum concentrations on the Earth are rather low among essential transition metals. On the other hand, it cannot be ignored that molybdenum is an abundant transition element in seawater. Several investigators questioned the 'direct panspermia' theory, pointing out the richness of molybdenum in seawater (Chappell et al., 1974; Jukes, 1974).

Jukes (1974) points out that molybdenum is more abundant in seawater than chromium or nickel, which does not support the panspermia hypothesis, as proposed by Crick and Orgel. Since molybdenum occurs in seawater in higher concentrations than nickel or chromium, it is necessary to postulate that the Mo stars might have served as a jumping-off point for panspermatic organisms that "infected the Earth".

According to Kobayashi and Ponnamperuma (1985) the role of trace elements as catalysts in chemical evolution may be summarized as follows: It is reasonable to assume that in the first stage of chemical evolution the fundamental bioorganic monomers (amino acids, nucleic acid bases, sugars, etc.) were formed from primitive atmospheres and seawater via the mediation of simple organic molecules such as hydrogen cyanide, formaldehyde and urea, using energy sources such as electric discharges, thermal energy, light, and shock waves. Several workers have reported that amino acids could be synthesized from simple molecules with the aid of trace metals. Bahadur et al. detected several amino acids after a paraformaldehyde solution was exposed to a tungsten lamp for 600 h with molybdenum oxide as a catalyst (Bahadur et al., 1958). Bahadur also reported that the solution of paraformaldehyde, potassium nitrate and ferric chloride, using sunlight as a source of energy, yielded about a dozen amino acids (Bahadur, 1954).

Oro et al. reported that various kinds of amino acids were produced after heating paraformaldehyde and hydroxylamine solution and that molybdic or vanadic acid could increase their yields (Oro et al., 1959).

Kobayashi and Ponnamperuma studied the synthesis of amino acids in electric discharge reactions in systems of methane, nitrogen, and ammonium buffer with and without metal ions. The system with metal ions (Mg^{2+}, Ca^{2+}, Zn^{2+}, Fe^{2+}, and $MoO_4{}^{2-}$) gave more amino acids than that without metal ions (Kobayashi and Ponnamperuma, 1985).

Many investigators have shown that some trace metal ions are able to catalyze the formation of some amino acids (such as glycine and alanine). However, other amino acids such as histidine and tyrosine have not been reported to be synthesized abiogenetically, with or without metal ions.

There have also been many interesting results on the roles of trace elements in the formation of nucleosides, nucleotides, and oligonucleotides. These studies show the importance of trace elements in abiotic formation of nucleic acid quite dramatically. Further studies are required on their role under more plausible prebiotic Earth conditions. In addition, the chemical forms of trace elements must be considered.

Calvin was one of the first to note the problem of evolution of enzymatic activity. He explained the evolution of a catalyst with peroxidase activity as an example as follows (Calvin, 1962). Aqueous ferric ion has a peroxidase activity, but at only 10^{-5} ml s^{-1} at $0^{o}C$. When ferric ion is surrounded by a porphyrin as in haem, its activity is increased 10^3 times. When the haem and the protein are bound as in 'catalase', its activity increases to as much as 10^5 ml s^{-1}. Porphyrin can be synthesized easily by electric discharge reactions (Hodgson and Ponnamperuma, 1968). Ferric iron and/or iron-porphyrin complexes can act as catalysts to synthesize porphyrin itself. Once an iron-porphyrin complex was manufactured in primitive seawater, it would accelerate its own synthesis from precursors; in other words, enzymes were developed autocatalytically. This autocatalytic feedback is the basic prerequisite of self-organization (Decker, 1981).

8. POTASSIUM/RUBIDIUM RATIO

Abundance of rubidium in all media is rather high. In seawater, its concentration is higher than the concentration of Fe-group transition metals. In soil, rubidium is an order of magnitude more abundant than Se, Br and Mo. In the universal abundance curve of Suess and Urey (1956) Rb is more abundant than Mo, with approximately the same abundance as Zn and As.

Rubidium is a member of the series NH_4-K-Rb-Cs. The members of this series are more alike in their chemical and physical properties than are the members of any other group, with the exception of halogens. The radius of the Rb ion is 1.48 Å (only about 10% larger than potassium ion radius). As the result, the Rb ion can be accommodated into the same structures as the K ion. Therefore, Rb does not even form the minerals of its own.

In spite of all this, rubidium seems not to be an essential trace element for life. It is present in almost all living material, but with K/Rb concentration ratios significantly increased than in inorganic material.

In complex biological systems there seem to exist close metabolic interrelationships between Rb and K. At least in humans, it was shown (Meltzer et al., 1969) that the excretion of each ion ought to be related to the other. A similar conclusion emerges after a broad survey of the literature. Rb, like K, is concentrated intracellularly. There is, however, a general tendency for Rb to have higher intracellular-extracellular ratios than K and it has been proposed that this indicates intracellular binding (Menozzi et al., 1959).

In some cases Rb can substitute for the K requirement (Muller, 1965). Where K is required for survival and growth, a complete replacement of dietary K by Rb allows normal growth for several weeks. Eventually, however, toxic effects appear. Lower chronic doses of Rb are not toxic.

Selectivity of Rb vs K intake has been studied in some animals (Love et al., 1954) and man (Kilpatrick et al., 1956). The ratios of radioactively labeled Rb and K in the blood plasma divided by the ratio in the urine, can be taken as an indicator of the selectivity of the renal tubule cells for Rb with respect to K. This ratio, which was found to be between 0.6 and 0.8 for the species studied, suggests that the renal tubule cells secrete less and/or absorb proportionately more Rb than K. These observations are summarized algebraically by the expression

$$Rb/K_{urine} = S \cdot (Rb/K)_{plasma}$$

where S is the selectivity factor denoting the preference of the renal tubule cells for Rb with respect to K.

It seems that the K/Rb ratio holds some clue on the functioning of living material. Even without understanding completely the mechanisms of the K and Rb differentiation by living cells the value of K/Rb ratios could be used in the search of extraterrestrial life forms. Although it is difficult to present K/Rb ratios for different materials because of possible systematic analytical errors, it seems that the K/Rb ratio is significantly higher for living matter than for inorganic material.

9. GEOMAGNETIC FIELD REVERSALS

While the distribution of essential elements in the periodic table reflects the element abundance in the environment in which life originated, the concentration factors and their dependance of physical and chemical parameters should reflect other properties of this environment. One such characteristic of the environment is the presence of magnetic field. Let us illustrate the possible approach to this problem on the example of geomagnetic field reversals.

In the distant past massive faunal extinctions have occurred. For example, roughly one-third of all living species became extinct at the close of the Cretaceous, which was a period marked by a resumption of polarity reversals of the Earth's magnetic field following a very lengthy period of normal magnetic activity. There is mounting evidence that a correlation exists between major faunal extinctions and geomagnetic polarity reversals. The validity of this correlation in recent geological time seems to have been well established by studies of fossil species of single-celled marine microorganisms.

Several mechanisms linking changes in the geomagnetic field with effects on living organisms have been proposed. Most of them are based on the assumption that during polarity reversals the dipole component of the geomagnetic field probably weakens or disappears for periods of a few thousands of years allowing a much greater flux of both solar protons and galactic cosmic rays to bombard the surface of the Earth. Other mechanisms include climate change, and the effects of a large reduction in the content of ozone in the atmosphere, which would increase exposure to ultraviolent radiation. Direct magnetic field effects on growth were discussed by Hays (1971) and Crain (1971).

A mechanism through which the geomagnetic field might have influenced the extinction of the species has been proposed by Valkovic (1977). In order to explain the geomagnetic effects on living organisms the concentration factor dependence on magnetic field intensity was assumed. During polarity reversals geomagnetic field intensity decreases by an order of magnitude. During this time species are affected by a magnetic field intensity of the order of magnitude of H_1. If they were living in an environment which had provided them with elements, they were in the range of adequacy when the Earth's magnetic field was H_2. Because of the assumed dependence of concentration factor on magnetic field intensity, they will be in the range of deficiency or toxicity when the magnetic field is H_1 (depending on the slope of the assumed functional dependence). Prolonged deficiency or toxicity for several generations may lead to disastrous effects including the species' disappearance. This can happen for one or more essential elements. Magnetic field dependence of concentration factors can bring living organisms into the range of deficiency or toxicity without changing trace element availability in the environment. One can imagine a species living in an environment such that a supply of a given essential trace element is near the edge of range of adequacy. In this case only small disturbances are needed in order to have an inadequate supply of the essential trace element and consequent development of abnormalities.

10. OPTICAL ACTIVITY

Large organic macromolecules, constituents of living cells such as proteins, are composed of amino acids which are able to rotate the polarization plane of light. Each optically active amino acid molecule can exist in two enantiomeric forms, the so-called L-form and D-form. These two forms are mirror images of each other and laboratory synthesis results in a racemic (50:50), optically inactive mixture. However, only L-amino acids are found in proteins in living organisms. Biological systems are based on L-amino acids and D-sugars.

Several asymmetric factors in the environment have been assumed to have had influenced this choice when life began. The suggestions included:

- polarized light,
- optically active quartz,
- natural radioactivity.

The present data about the origin of life state that micro organisms such as bacteria and blue-green algae existed 3.2-3.3 x 10^9 years ago. The oldest dated rocks are 3.8 x 10^9 years old, the chemical evolution of biomolecules on the Earth is also assumed to have originated around this period. Therefore, it is often assumed that optical activity - assuming that it was induced by natural radioactivity - have had to evolve during the chemical evolution, i.e., in the time interval 3.3-3.8 x 10^9 years ago (Keszthelyi, 1976) if the optical activity originated on the Earth.

In this scenario the amino acids were synthesized mainly in the atmosphere, while the β^--radioactive nuclei (members of U and Th decay series amd ^{40}K) were found in the Earth's crust and in the sea.

All the proposed theories for the origin of optical activity are summarized in Fig. 10.

As we have already stated, the elemental synthesis takes place in stars. After several cycles of burning-contraction processes, the star

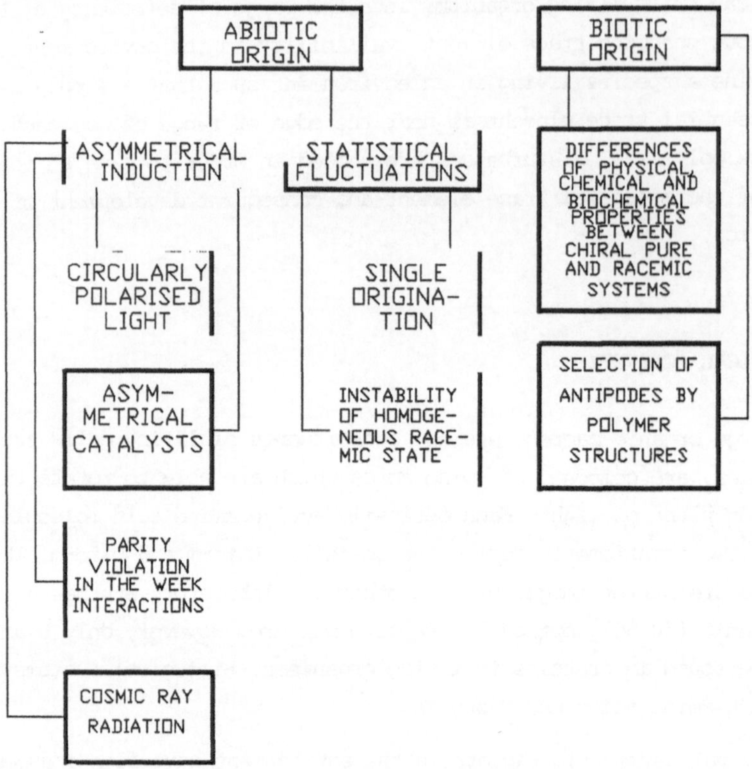

Fig. 10. Proposed theories for the origin of optical activity.

must die. This often happens by explosions, producing dust particles. During the explosions, final abundances of elements are determined. Inspection of the distribution of the essential elements in the periodic table of elements shows that this distribution reflects the properties of nuclear reaction chains. For example, elements not made in the pp chain are not essential for life, although these elements are abundant in the terrestrial environment - soil and water. Starting from this, the assumptions can be made that the life originated on dust particles, the universal building blocks of new stars and their planetary systems. This assumption may be helpful in explaining the origin of the optical activity of sugars and amino acids. If life originated on dust particles, then the most probable factor responsible for the predominance of only one optical isomer is the exposure to cosmic ray bombardment. The preferential destruction of one of the isomers may result from the interaction of polarized protons with hydrogen nuclei in the molecules. The observed asymmetries in the cross section for p-p elastic scattering in a wide range of proton energies support this assumption. High energy polarized protons in cosmic rays may be able to preferentially destroy one isomer because of large asymmetry in proton (in cosmic ray) - proton (in amino acid) scattering.

11. INTERSTELLAR DUST

Dust is associated with gas clouds. In the neighbourhood of the Sun clouds are collected mainly in the arms of the Galaxy in which the Sun is situated.

The grains of interstellar matter are formed by molecular collisions. The starting point would be the formation of CH molecules by $C + H \longrightarrow CH$. From CH molecules, higher order hydrocarbons or chains of carbon-metal-carbon might be formed. These big molecules would form a closed ring when they contain 50 - 200 atoms. From that moment these chains of atoms are capable of sweeping up matter in the interstellar space. The growth of particles could be rapid (0.1 micrometer per million years). Evaporation from grain surfaces limits the growth of particles.

The observed polarization of light is explained by the presence of anisotropic particles suitably oriented by magnetic field directed along the spiral arms.

Studies of the depletion of interstellar elements from the gas phase provide an indirect method for the investigation of the interstellar dust. The large depletions for Si, Mg, Fe, Cr, Ti, Ca, Ni, and Al are observed and imply that these species are nearly completely tied up in the dust. In contrast, S and Zn are not greatly depleted.

It was suggested that the composition of dust grains must change with the decreasing distance from the galactic center. This supports the existence of an abundance gradient of heavy elements in our Galaxy, as it has been observed in external galaxies (Searle, 1971; Churchwell et al., 1974).

There has been speculation about interstellar grains being composed of organic substances (bacterial cells and viruses). According to Khare and Sagan (1979), a complex array of organic solids was probably produced in the primitive solar nebula and contributed to the present composition of carbonoceous chondrites and comets as well as to the interstellar grains and gas.

Interstellar molecular chemistry is discussed in details by many authors (see, for example, Turner, 1980). It is generally accepted that the chemistry of interstellar molecular regions appears to be very complex: the formation and destruction of molecules can involve atoms, ions, free radicals, molecular ions, neutral molecules, solid surfaces, cosmic rays, energy from shocks, and radiation fields.

The interstellar dust appears to play a critical role in the formation of the interstellar molecules. Molecules may be formed on or in grain surfaces. The existence of interstellar molecules suggests that:

(1) such molecules support or are the metabolic products of an interstellar biota;

(2) such molecules, participating in planetary condensation from the interstellar medium, can make a significant contribution to the origin of planetary life.

From all these considerations the following assumptions could be made:

- life is universal;
- life originated on dust particles, the universal building blocks of new stars and their planetary systems.

Many authors have considered the possibility and implications of comets being sources of volatiles, particularly H_2O, to the Earth's paleo-atmosphere. These processes could have involved interplanetary dust particles, which are mostly generated by gradual disintegration of short-period comets in the inner solar system. In the elemental composition they are similar to chondritic abundances. The similarity between interplanetary material and carbonaceous chondrites indicates that they are both well preserved samples of early solar system material. Small dust particles, because of their large surface to mass ratio, can enter the Earth's atmosphere without melting.

12. CONCLUSIONS

There are two possible approaches to the understanding of the origin of trace element requirements:

1. Either the choice of an essential trace element was made partly based on the availability of elements in the environment in which the life originated;
2. or, if the predetermined ratios of abundances of some selecting elements are required for the living matter functioning, the life orginated when in the development of universal abundance curve these ratios were reached.

This latter assumption would explain while we do not observe any more the process of the origins of life. In both cases the interstellar dust particles appear as possible sites of these processes.

REFERENCES

Bahadur, K., 1954, Nature, 173:1141.

Bahadur, K, Ranganayaki, S., and Santamaria, L., 1958, Nature, 182:1668.

Banin, A., and Navrot, J., 1975, Science, 189:550.

Bowen, H. J. M., 1966, "Trace Elements in Biochemistry", Academic Press, New York.

Calvin, M., 1962, Perspect. Biol. Med., 5:399.

Chappell, W. R., Meglen, R. R., and Runnells, D. D., 1974, 18:513.

Churchwell, E., Mezger, P. G., Huchtmeier, W., 1974, Astron. and Astrophys. 32-283.

Cotzias, G. C., and Foradori, A. C., 1969, in "The Biological Basis of Medicine" (Bittar, E. E. and Bittar, N., Eds.), Academic Press, New York.

Crain, I. K., 1971, Bull. Geol. Soc. Am., 82:2603.

Crick, F. G. C., and Orgel, L.E., 1973, Ixarus, 19:341.

Decker, P., 1981, in "Origin of Life" (Y. Wolman, Ed.), D. Reidel, Dordrecht, Holland, pp. 529.

Egami, F., 1974, J. Mol. Evol., 4:113.

Fisher, N. S., 1986, Limnol. Oceanogr., 31:443.

Fisher, N. S., and Fowler, S. W., 1987, in "Oceanic Process in Marine Pollution" (O'Connor, T. P., Burt, W. V., and Duedall, I. W., Eds.), Robert E. Krieger Publ. Co., Malabar, Florida.

Hays, J. D., 1971, Bull. Geol. Soc. Am., 82:2433.

Hodgson, G. W., and Ponnamperuma, C., 1968, Proc. Nat. Acad. Sci. USA. 59-22.

Johnson, F. M., 1972, Ann. N. Y. Acad. Sci., 187:186.

Jukes, T. H., 1974, 21:516.

Keszthelyi, L., 1976, Origins of Life, 7:349.

Khare, B. N., and Sagan, C., 1979, Astrophysics and Space Sciences, 65:309.

Kilpatrick, R., Renschler, H. E., Munro, D. S., and Wilson, G. M., 1956, J. Physiol., 133:194.

Kobayashi, K., and Ponnamperuma, C., 1985, Origins of Life, 16:41.

Krasnovsky, A. A., 1976, Origins of Life, 7:133.

Love, W. D., Romney, R. B., and Burch, G. E., 1954, Circ. Res., 2:112.

McClendon, J. H., 1976, J. Mol. Evol., 8:175.

Meltzer, H. L., Taylor, R. M., Platman, S. R., and Fieve, R. R., 1969, Nature, 223:321.

Mennozzi, P., Norman, D., Polleri, A., Lester, G., and Hechter, O., 1959, Proc. Nat. Acad. Sci. USA, 45:80.

Muller, P., 1965, J. Physiol., 117:453.

Orgel, L. E., 1974, Icarus, 21:518.

Oro, J., Kimball, A., Fritz, R., and Mater, F., 1959, Arch. Biochem. Biophys., 85:115.

Parker, R., 1967, U.S. Geol. Survey Prof. Paper, 440-D.

Searle, L., 1971, Astrophys. J., 168:327.

Steenbjerg, F., 1951, Plant Soil, 3:97.

Suess, H., and Urey, H. C., 1956, Rev. Mod. Phys., 28:53.

Turner, B. E., 1980, J. Mol. Evol., 15:79.

Wacker, W. E. C., and Vallee, B. L., 1959, J. Biol. Chem., 234:3257.

Valkovic, V., 1977, Origins of Life, 8:7.

Valkovic, V., 1980, Origins of Life, 10:301.

Valkovic, V., 1980, "Analysis of Biological Material for Trace Elements Using X-Ray Spectroscopy", CRC Press, Boca Raton, Florida.

INDEX